Quantitative and Qualitative Determination Technologies of Counterfeit Drugs

Drugs are often counterfeited to reduce manufacture costs, while still marketing it at as an authentic product. Increased incidence of drug counterfeiting is most noticeable in developing countries, which may not have the resources to supply counterfeit detection devices on a large scale. It is important to consider the problems caused and to propose options for controlling and reducing the prevalence of counterfeit medications. Various technologies are needed to identify the chemical properties of a questioned medicinal product, which can then be used to determine its authenticity. This volume focuses on current technological approaches that are able to detect counterfeited pharmaceuticals.

Features

- Focuses on current technological approaches that are able to detect counterfeited pharmaceuticals.
- Assesses the chemical methods of identifying counterfeit medicinal products and explains the theoretical underpinnings of the methods.
- Provides case study type examples of the application for analysis of suspected counterfeit pharmaceuticals.
- Discusses the detection and analysis of counterfeit drugs, and appropriate tools for combating this issue.
- The editor draws on his experience as a respected chemist and prolific author in the field to provide this unique text on drug counterfeiting detection.

Counterfeit Drugs Series
Series Editor
Kelly M. Elkins

Trends in Counterfeit Drugs
Edited by Kelly M. Elkins

Quantitative and Qualitative Determination Technologies on Counterfeit Drugs
Edited by Ronny Priefer

Quantitative and Qualitative Determination Technologies of Counterfeit Drugs

Edited By
Ronny Priefer

CRC Press
Taylor & Francis Group
Boca Raton London New York

CRC Press is an imprint of the
Taylor & Francis Group, an **informa** business

First edition published 2024
by CRC Press
6000 Broken Sound Parkway NW, Suite 300, Boca Raton, FL 33487-2742

and by CRC Press
4 Park Square, Milton Park, Abingdon, Oxon, OX14 4RN

CRC Press is an imprint of Taylor & Francis Group, LLC

Library of Congress Cataloging-in-Publication Data
Names: Priefer, Ronny, editor.
Title: Quantitative and qualitative determination technologies of counterfeit drugs /
edited by Ronny Priefer. Other titles: Counterfeit drugs series.
Description: First edition. | Boca Raton : CRC Press, 2023. | Series: Counterfeit drugs series |
Includes bibliographical references and index. |
Identifiers: LCCN 2023003286 (print) | LCCN 2023003287 (ebook) |
ISBN 9781032218922 (hardback) | ISBN 9781032218984 (paperback) |
ISBN 9781003270461 (ebook)
Subjects: MESH: Counterfeit Drugs–analysis | Drug and Narcotic Control |
Technology, Pharmaceutical | Fraud–prevention & control Classification:
LCC RS189 (print) | LCC RS189 (ebook) | NLM QV 773 |
DDC 338.476153–dc23/eng/20230516
LC record available at https://lccn.loc.gov/2023003286
LC ebook record available at https://lccn.loc.gov/2023003287

ISBN: 978-1-032-21892-2 (HB)
ISBN: 978-1-032-21898-4 (PB)
ISBN: 978-1-003-27046-1 (EB)

DOI: 10.1201/9781003270461

Typeset in Times
by Newgen Publishing UK

Contents

Preface vii

Editor ix

Contributors xi

Chapter 1 Screening for Bad-Quality Pharmaceuticals in Field Settings 1

Marya Lieberman

Chapter 2 HPLC for Analysis of Substandard and Falsified
Medicinal Products 39

Eric Deconinck and Celine Vanhee

Chapter 3 Tackling a Global Crisis: A Review of the Efficacy of LC-MS 77

Dong Hyeon Shin and Ronny Priefer

Chapter 4 Analyzing Counterfeit and Falsified Herbal Products
with GC/MS 97

Maryam Akhgari and Afshar Etemadi-Aleagha

Chapter 5 Applications of UV-Vis Spectroscopy in Counterfeit
Medications 119

Jin Ba and Ronny Priefer

Chapter 6 IR Spectroscopic Analytical Tools in the Fight Against
Counterfeit Medicines 131

Sangeeta Tanna and Rachel Armitage

Chapter 7 Using NMR Spectroscopy to Analyze Counterfeit
Medications 171

Alina Hoxha and Ronny Priefer

Index 193

Preface

Although food and drug regulatory agencies exist that work diligently to ensure that the products that enter the marketplace are of the quality and quantity that they were approved at; whether accidently or maliciously, all too often counterfeits are identified. This issue has seen a noticeable decrease in developed nations due to numerous checks and balances, however cases are still periodically reported. More troubling is the astounding level of falsified drugs that are in developing countries, particular throughout Africa. For example, studies have shown that approximately 15% of antimalarial and antibiotic agents in and around the sub-Sahara are either sub-therapeutic or completely devoid of any active pharmaceutical ingredients. Additionally, as more sales of drugs are transitioning to online purchasing, less oversight is being done, which allows for further increases in this arena.

To combat issues related to counterfeit pharmaceuticals, a variety of different analytical techniques have been developed. These range from the inexpensive PADs and chromatography approaches to incredibly precise, yet notoriously expensive, NMR and MS. In addition to the cost disparity, which negates application of many of these technologies in developing countries, some techniques cannot be utilized for certain drug formulations. A global group of experts collaborated to bring about a comprehensive overview of these technologies, highlighting each of their strengths and weaknesses. The information provided in this book will greatly assist those currently in the field attempting to combat this growing global problem, as well as scientists and engineers seeking to develop new technologies to reverse this trend.

Editor

Ronny Priefer earned his PhD from McGill University in Montreal, Canada in Organic Chemistry. He is a full professor and Dean of Graduate Studies in the School of Pharmacy at the Massachusetts College of Pharmacy and Health Sciences (MCPHS) University. Prior to this, he was a professor of medicinal chemistry at the College of Pharmacy at Western New England University and at Niagara University in the Chemistry and Biochemistry Department. His research areas have focused on novel surface coating in conjunction with drug delivery and stability. He has over 100 peer-reviewed publications and multiple patents.

Contributors

Maryam Akhgari
Legal Medicine Research Center, Legal medicine Organization, Tehran, Iran

Rachel Armitage
Faculty of Health and Life Sciences, De Montfort University, Leicester, United Kingdom

Jin Ba
MCPHS University, Boston, Massachusetts

Eric Deconinck
Scientific Direction Chemical and Physical Health Risks, Service of Medicines and Health Products, Brussels, Belgium

Afshar Etemadi-Aleagha
Tehran University of Medical Sciences Tehran, Iran

Alina Hoxha
MCPHS University, Boston, Massachusetts

Marya Lieberman
Department of Chemistry and Biochemistry, University of Notre Dame, Notre Dame, Indiana

Ronny Priefer
MCPHS University, Boston, Massachusetts

Dong Hyeon Shin
MCPHS University, Boston, Massachusetts

Sangeeta Tanna
Leicester School of Pharmacy, De Montfort University, Leicester, United Kingdom

Celine Vanhee
Scientific Direction Chemical and Physical Health Risks, Service of Medicines and Health Products, Brussels, Belgium

1 Screening for Bad-Quality Pharmaceuticals in Field Settings

Marya Lieberman
Department of Chemistry and Biochemistry,
University of Notre Dame, Notre Dame, IN

CONTENTS

1.1 Introduction 2
   1.1.1 Scope and Scale of the Problem 2
   1.1.2 Who Defines Quality? 3
   1.1.3 What do Regulators in LMICs do to Help Catch SFPs? 4
   1.1.4 What is Field Screening? 6
   1.1.5 How Does Field Screening Help Solve the Problem of SFPs? 6
1.2 What can go wrong with pharmaceuticals? 7
   1.2.1 Pharmaceuticals Can be Badly Made 7
   1.2.2 Pharmaceuticals can be Badly Packaged or Transported
        and Stored in Ways that Cause Degradation of Quality 9
   1.2.3 Pharmaceuticals Can Be Dispensed Badly 10
   1.2.4 The Manufacturer Can Be Unresponsive 11
1.3 Instrument-Free Field Screening Technologies 11
   1.3.1 What Analytical Metrics are Appropriate for Field Screening
        Technologies? 11
   1.3.2 Packaging Inspection 12
   1.3.3 Glassware-Based Field Color Tests 13
   1.3.4 Thin-Layer Chromatography 14
   1.3.5 Merck-GPHF Minilab™ 15
   1.3.6 Disintegration/Dissolution Testing 18
   1.3.7 Lateral Flow Immunoassay Strips 18
   1.3.8 Paper Microfluidics: the Paper Analytical Device (PAD) 19
        1.3.8.1 How the PAD Works 19
        1.3.8.2 Test Development 20
        1.3.8.3 Making PADs 20
        1.3.8.4 Data Analytics for the PAD 21
1.4 An Implementation Case Study: Paper Analytical Device (PAD) 23

DOI: 10.1201/9781003270461-1

1.5   Who Tests the Tests? 25
        1.5.1   LOMWRU Multiphase Review..26
        1.5.2   USP Technology Review Program...27
1.6   Conclusion...29
References...30

1.1   INTRODUCTION

Buying a pharmaceutical is not like buying fruit at the market. A banana is there in your hands; if it is rotten, you can see, smell, and feel the quality problem. A capsule of amoxicillin comes hidden in its blister packaging or in a pill bottle – you cannot sniff it or check the color of the powder – and if you do open a capsule and pour out some white powder, who's to know if it is amoxicillin or ground chalk? A person at a market can make an informed decision about whether to put that banana in their basket; a patient must trust that the pharmaceutical is what it claims to be when they put it in their mouth. According to a WHO meta-analysis in high-income countries, this trust is almost always justified. In low- and middle-income countries, about one times in ten, the pharmaceutical will be substandard or falsified.[1]

1.1.1   SCOPE AND SCALE OF THE PROBLEM

Substandard and falsified pharmaceuticals (SFPs) are not a new problem. In a manuscript penned before 70 CE, Dioscorides describes how to distinguish genuine opium from fakes concocted from the juice of wild lettuce.[2] In 1868, J. Broughton reports a herbal product sold for treatment of malaria, which has the appearance and bitter taste of cinchona bark but contains the toxic compound esculin in place of the antimalarial quinine.[3] The World Health Organization's 22nd Essential Medicines List contains hundreds of pharmaceuticals, from anesthetics to vitamins, deemed to be critical for a functional medical system. In 2017, a WHO literature survey found examples of substandard or falsified versions for every class of these pharmaceuticals.[4]

The WHO survey estimated that the prevalence of SFPs across pharmaceuticals in low- and middle-income countries (LMICs) was about 10%, much higher than in high-income countries. The combination of buyers who are strongly price driven, bad actors in the supply chain, and lack of regulatory resources all contribute to the continuing high levels of SFPs in the medical systems of LMICs.[5] Despite the evidence for widespread and harmful failure of quality assurance systems,[6] there is still no overarching authority that conducts systematic post-market surveillance (PMS) of essential pharmaceutical products at a global scale.[7] In most LMICs, this vital function is conducted by poorly resourced national drug regulatory agencies leading to piecemeal, reactive, and ineffective efforts to keep bad-quality products out of the market.[8,9]

1.1.2   Who Defines Quality?

Pharmaceutical quality is defined by **pharmacopeial organizations**. There are 21 active pharmacopeial organizations that publish monographs containing specifications for pharmaceutical products under the authority of a government or pharmaceutical society. Some pharmacopeias are supranational, such as the International Pharmacopeia (IP), published by the WHO, while others are national, such as the US Pharmacopeial Convention (USP). In addition to providing standards for finished pharmaceutical products, pharmacopeias provide standards for active pharmaceutical ingredients (APIs) and excipients. These standards are expressed in product monographs and include detailed descriptions of the quality standards and the methods of analysis to be used to verify whether the product meets each quality standard. The analytical methods that determine whether a product passes or fails the quality standards are called **compendial methods**. Products that meet all the monograph standards are good quality by definition. For example, under the USP's monograph for ampicillin capsules,[10] there are references to simple identity tests based on color reactions and infrared spectroscopy to confirm the presence of the expected API; the ampicillin content is required to be within a range of 90%–120% of the stated content and a high performance liquid chromatography (HPLC) assay method and titration-based assay method are specified. Additional standards are given for packaging, labeling, and performance attributes of the ampicillin capsules, to be measured through dissolution testing, content uniformity, and water content. Full monograph testing of a product by compendial methods must take place in a well-equipped and managed laboratory with trained personnel. It is expensive and time consuming.

Many literature reports of SFPs are based on partial monograph testing or non-compendial analysis methods, including results from tests that are regarded as field screening tests. Depending on how the enabling legislation is written, non-monograph analysis methods may be unacceptable for purposes of regulatory compliance or enforcement, but performing these tests can rapidly uncover problem products or batches that can then be subjected to compendial testing.

The product specifications and analysis methods written by different pharmacopeial organizations do not always agree. In the British Pharmacopeia monograph for ampicillin capsules, the ampicillin content is required to be in the range of 90%–110%; the USP monograph sets the acceptable range as 90%–120%. Thus, a product whose capsules contain 115% of the stated ampicillin content is substandard according to BP, but meets standard according to USP. The quantity units used for medicines are not globally consistent; for example, the API content of an antibiotic that is formulated as a citrate or chloride salt may be calculated as milligrams of the salt or milligrams of the free base, and amounts of other medicines, such as oxytocin, may be expressed in terms of international units (IU) rather than as masses. Drug regulatory agencies of the US, Japan, and Europe have been meeting for over 40 years to attempt to iron out such inconsistencies.

Monograph methods start by defining the physical form of the product and the active pharmaceutical ingredient (API). Some APIs include water of crystallization, others can be present as free acids or bases or as salts, and some are defined in various units based on biological activity. For small-molecule drugs that are taken orally, monograph methods include identification, quantitative assay, and performance tests. Identification methods typically include infrared (IR) spectroscopy, chemical color tests, and/or thin-layer chromatography (TLC). In each case, the results must be compared to results from a reference standard of known good quality and a clear criterion for acceptance is provided. Monograph methods for the quantitative assay of APIs are dominated by high performance liquid chromatography (HPLC) analysis using optical absorption in the ultraviolet/visible (UV-Vis) range for detection of peaks. These methods require high quality reference materials, high purity solvents and buffer components, and accurate massing of sample and reference materials. There are a few other quantitative monograph methods, including liquid chromatography with mass spectrometric detection (LCMS), gas chromatography (GC), UV-Vis, and titrimetric assay. Monograph assays are usually performed on pooled contents of multiple dosage forms, usually requiring more dosage units than are provided in a course of treatment for one patient. Performance of dosage forms includes assessing the content uniformity and water content and verifying that oral medicines dissolve correctly so the medicine is bioavailable. The dissolution assay requires a specialized instrument and capacity to assay the amount of the API released into solution over time.

Preparation of monograph methods is time consuming and difficult, so many governments in LMICs instead adopt pharmacopeial monographs and other regulatory standards used by one of the stringent regulatory authorities or by a supranational organization like the International Pharmacopeia. However, when the monographs are written by organizations that are based in high-income countries (HICs), monograph methods may not be provided for medicines that are primarily used in LMICs. Roth et al. reviewed availability of monograph methods for 669 essential medicines in 2017, evaluating in eight major pharmacopeias, and found that 15% of the medicines did not correspond to any available monographs.[11] This included 28% (9/32) of the antiretroviral medicines and 23% (6/26) of the antimalarial medicines.

1.1.3 What do Regulators in LMICs do to Help Catch SFPs?

Because the problem of SFPs is concentrated in LMICs and global enforcement activities are relatively weak, regulatory agencies in LMICs are the frontline organizations responsible for exposing and controlling SFPs. Regulatory activities in LMICs include registration, import/export controls, post-market surveillance (PMS), and enforcement.[12]

Registration is the process by which companies obtain approval to sell a particular product in the country. It requires submission of a dossier of information about the product and its manufacturer, documenting both the efficacy of the medicine and the ability of the manufacturer to produce a good quality product. Many regulatory agencies require submission of samples of the product for chemical analysis in a medicine quality control laboratory at the time of registration. The supplier pays for this lab testing and depends on a satisfactory result for approval of the dossier, so it is not surprising that products submitted for registration testing generally pass. Testing of registration samples in LMIC medicine quality control labs is often a major source of their funding. This creates unintended consequences: it ties up the laboratory in analysis of samples that are likely to be of good quality, and it creates an incentive to register hundreds of brands of the same basic medicines. All of these different products must then be monitored in the future.

Import and export controls provide some follow-up information about which products are entering or leaving the country. However, many LMIC regulators only use paperwork controls to check product quality, for example requiring submission of supplier-provided lab results for new batches at the point of import. A manufacturer or distributor that is trying to sell an SFP could also falsify lab results and import paperwork.

Post-market surveillance (PMS) probes the quality of products that are actually out in the market, which makes it a critical measure of the quality of drugs that patients are exposed to. Unlike registration and import testing, PMS can detect problems in drugs that have been brought into the country illicitly or through regulatory loopholes, such as compassionate use exemptions or charitable organization supply chains. PMS can be conducted either overtly or covertly, using convenience samples or random samples, and with a wide range of analysis methods. Several guidelines have been developed [13] and there are WHO recommendations for best practice.[14] The best type of PMS can identify which brands are present in the market at a particular point in time, their relative market shares, and the quality of these brands (Figure 1.1).

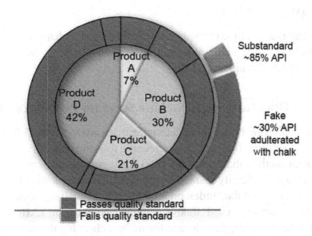

FIGURE 1.1 Post-market surveillance helps to prioritize regulatory response. In this market, product B presents the most risk to patients. Photo: Marya Lieberman.

PMS is limited by access to laboratory assay of the collected samples. The number and types of products that are collected in a PMS campaign are often focused to meet a particular health need or follow up on a past SFP or an SFP that has been reported in a neighboring country. A PMS project might involve collecting and testing 300 antibiotic products. This sounds like a lot of samples, but if a country has registered 102 brands of ciprofloxacin, which are imported by four major distributors and sold in multiple regions of the country, the coverage of this PMS campaign will actually be insufficient to test each brand of ciprofloxacin, let alone all the other types of antibiotics used in the country.

Enforcement activities may include issuing notices, regulatory controls such as quarantining a product (requiring sellers to set it aside while lab testing is going on) or recalling it, banning the supplier from selling products in the country, or initiating legal actions against the supplier. The financial consequences of enforcement activities are heavy and may be borne by small medicine shop owners or medicine distributors rather than manufacturers, particularly when the medicine in question is not registered properly or when the manufacturer cannot be identified or does not cooperate with the recall.

Enforcement activities in most LMICs do not catch most SF products in time to prevent harm to patients, as witness the 10% rate of SF products in these markets. The market for many pharmaceuticals in LMICs is very lively, with rapid turnover of products on a week to month time scale. Post-market surveillance often takes more than a year between collection of the samples and issuance of reports or enforcement actions. Thus, by the time a problem product has been discovered, it has already been sold to many patients. If the product is not in the market any more, issuing recalls, bans, or quarantines on it are like closing the barn door after the cow has been stolen.

1.1.4   WHAT IS FIELD SCREENING?

Field screening is a tentative evaluation of the quality of a pharmaceutical product and is conducted outside a laboratory setting. It is widely recognized as a valuable part of post-market surveillance.[15]

1.1.5   HOW DOES FIELD SCREENING HELP SOLVE THE PROBLEM OF SFPS?

Most field screening methods cannot determine that a product is an SFP, because that is a regulatory determination, and field screening tests are generally not included in the compendial method that forms the basis for regulatory action. However, field screening can rapidly identify suspicious products for further testing by compendial methods, which reduces the burden on pharmaceutical analysis laboratories.

The most commonly used field screening methods in LMICs are visual inspection of the packaging and dosage forms, and checking the written expiration dates and registration status of the product. These procedures offer clear-cut criteria for whether a sample passes or falls. Recently, packaging security features such as holographic stickers, barcodes for track-and-trace systems, and scratch-off codes have become more common. These features are easy to use but they provide security for the packaging, rather than the drugs inside; see below.

Regulatory agency respondents in nine out of ten countries surveyed by Roth *et al.*, agreed with the cost-saving arguments for field screening of pharmaceuticals. [16] However, most of the respondents misunderstood the capabilities of current chemical and spectroscopic screening technologies. These technologies are developing rapidly and have great potential as weapons against SFPs, but there are still many evidence gaps and it is difficult for regulators to compare the effectiveness, usability, and costs of different technologies.

1.2　WHAT CAN GO WRONG WITH PHARMACEUTICALS?

A good starting point in evaluating field screening technologies is to consider the questions that regulators are trying to answer when they assess whether a pharmaceutical is good quality or not. These questions are directly related to the many ways that the manufacture, distribution, storage, and dispensation of pharmaceuticals can go wrong. This section focuses on small molecule drugs, not biologicals or vaccines, and will not cover the important topic of assuring sterility in injectable formulations.

The bulk of the products in the WHO Essential Medicines List are oral or injected drugs, almost all of which are small molecule drugs that are off patent, thus available for manufacture by multiple companies.[17] Drug formulations include an active pharmaceutical ingredient (API) or sometimes multiple APIs, along with excipients, which are materials added to improve the bioavailability, stability, and performance of the formulation. Oral formulations include capsules (the API and a small quantity of anti-caking and other excipients are placed into a gelatin capsule), tablets (tableting excipients and the API are pressed into a tablet, which may be coated to aid swallowing), and syrups or suspensions. Injectable formulations may be prepared as pre-made solutions or as solids to be reconstituted by addition of sterile diluent; these are usually sold in small septum-capped vials.

Pharmaceutical manufacturers must follow good manufacturing practices (GMP) which include high levels of cleanliness and safety, reproducibility of processes, documentation, and QA/QC. Pharmaceuticals must be stored, shipped, and dispensed following good distribution practices, which include controlling temperatures and moisture, time in transit, and documentation. When these practices are not adhered to, the outcome is often an SFP.

1.2.1　PHARMACEUTICALS CAN BE BADLY MADE

Some SFPs are outright fakes, purposefully made and sold with lethal contempt for the people using the product. A manufacturer (still unidentified and unpunished) made counterfeit Coartem tablets of starch and yellow food coloring; these tablets are nearly indistinguishable from the authentic product to the eye (Figure 1.2), and

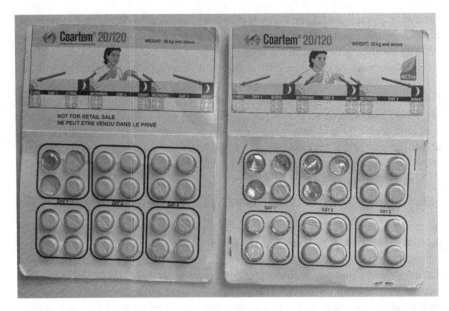

FIGURE 1.2   Real (left) and falsified (right) Coartem. The staples on the fake package are a giveaway. Later versions of this falsified product used glue (the real product uses a pressure-sensitive adhesive film). Photo: Marya Lieberman.

they are packaged in a convincing version of the authentic blister package.[18] Researchers estimate that ineffective antimalarials like this product are responsible for over 120,000 malaria deaths each year in children under five.[19]

Legitimate pharmaceutical manufacturers can also produce drugs that fail quality standards. Manufacturers purchase ingredients, such as APIs, from other manufacturers, and if they do not have adequate quality testing practices, they may incorporate a poor quality or fake API or excipient into their product.[20] Poor control over manufacturing processes can also lead to production of SFPs. In 2010, a small pharmaceutical manufacturer in Lahore, Pakistan received a shipment of pyrimethamine; the bags of this API were transferred into bins that were normally used to hold starch, and during manufacture of "Isotab" isosorbide mononitrate tablets, the pyrimethamine was mistakenly added instead of starch during the tableting process. Heart patients who took this batch of Isotab were exposed to such high doses of pyrimethamine that it damaged their bone marrow; more than 200 people died. This was not a "counterfeit" product – the manufacturer was the legitimate producer with the proper registrations – but an example of negligence so severe the Pakistani judiciary called it "criminal".[21]

The willingness of some pharmaceutical companies to produce both good-quality products (for stringent regulatory markets) and intentionally substandard products (for weak regulatory markets) has been well documented.[22] However, because counterfeit products can be packaged in good imitations of a legitimate manufacturer's packaging materials, it can be difficult to distinguish between a company that deliberately produces a bad-quality product, and a company that is the victim of deceptive counterfeiting (bad-quality products made by another manufacturer that imitate good-quality products).

1.2.2 PHARMACEUTICALS CAN BE BADLY PACKAGED OR TRANSPORTED AND STORED IN WAYS THAT CAUSE DEGRADATION OF QUALITY

Pharmaceutical standards include detailed directions for how the product should be packaged after formulation. The packaging materials serve multiple functions. First, they act as barriers to protect the formulated product from moisture, air, and light, all of which can cause degradation of the active ingredients or physical deterioration of the formulated product. Second, they link the product to information useful to the patient, such as the product name, active ingredient, and dosage, directions for how often to take the medicine, and whether it should be taken with food. Third, they link the product to information useful to regulatory authorities, such as the manufacturer's name and address, product expiration date, and lot or batch number. Common types of pharmaceutical packaging include blister packaging and heat-sealed foil packaging for tablets or capsules and septum-capped vials for injectable drugs, often co-packaged with aseptically prepared diluent. These primary packaging units may be further packaged in sealed pouches or boxes along with informational leaflets.

Physical packaging inspection begins with assessing the condition of each layer of packaging, if possible comparing the packaging from multiple lot numbers of the same product. Evidence of water damage or mold may be visible on the box. The adhesive or seal on blister packaging and heat-sealed foil strips may be degraded or failed. The colors of the inks and other packaging materials, the shapes and sizes of the packaging elements, and the printed information are carefully compared to identify possible deceptive counterfeits of a legitimate product. The lot numbers and expiration date stamped on the primary packaging are compared to the lot numbers and expiration date printed on the box to ward against repackaging or re-dating. Finally, the tablets or capsules are examined to see if they have the expected shape, color, and imprint, or are broken, cracked, sticky, or damp. If a milligram balance is available, tablets or capsules may be weighed.

While gross physical problems with the dosage forms such as crumbling or foul-smelling tablets arouse legitimate concern about the quality of the API, the utility of packaging inspection as a guideline for pharmaceutical quality is weak. It is common to see multiple versions of packaging materials for a single brand in an LMIC market. Just as manufacturers shop around to get a better deal on a batch of an API, they also seek less expensive inks, glues, or heat-sealable foil for their packaging line. Some manufacturers appear to subcontract the formulation, tableting, and primary packaging to specialty manufacturers, receiving products in custom-printed blister packs that they then package for distribution. Packaging for different lot numbers of the same brand may include different ink colors and

Manufacturer: "the two samples were made in different manufacturing facilities and packaging lines, but both are authentic"

Manufacturer: "The batchnumber is an officially released batchnumber. The printing error is pointed out by our quality department."

FIGURE 1.3   Packaging differences found on authentic products. Photo: Marya Lieberman.

typographic fonts, different embossing patterns, or different types of paper and plastic.[23] Spelling errors or poorly translated text may be present on legitimate products. (Figure 1.3) Although several groups (including the WHO) have published guidelines for conducting packaging analysis, [24,25] the general effectiveness of physical packaging analysis for pharmaceutical screening purposes has not been reported.

1.2.3   PHARMACEUTICALS CAN BE DISPENSED BADLY

Good practices for pharmaceutical dispensing include checking the patient's prescription for prescription-only medicines, providing the patient with information about the product such as how to store and use the medicine, and, of course, providing the correct product. In many LMICs, prescription-only medicines such as diazepam are widely available without a prescription.[26] Repackaging of pharmaceuticals is also common practice, for example, capsules counted from a bulk jar may be placed in a paper or plastic envelope, or several pills may be cut from a blister-packed set. This practice complicates post-market surveillance because the information normally provided on the packaging (e.g. brand name, batch number) may be lost, and the patient may not receive the right number of pills for a full course of treatment.[27] In a clear step across the line of legality, products which have passed their expiration date may be repackaged and redated, and most concerningly, one drug may be sold as another. From December 2014 through August 2015, a mysterious outbreak of muscle spasms of the face, neck, or arm affected more than thousand children in the Democratic Republic of Congo. Meningitis was ruled out, and the outbreak was traced back to drug substitution by an in-country distributor. The children had been given what caregivers thought was diazepam, but the tablets were actually repackaged haloperidol tablets, which can cause severe dystonic symptoms.[28]

1.2.4 THE MANUFACTURER CAN BE UNRESPONSIVE

Although manufacturers are required to provide contact information for quality control purposes, it is common for these emails or phone numbers to be outdated, disconnected, or to lead to a messaging system or phone tree from which no response is ever received. Companies may go out of business, or they may be bought out by other companies. It is often difficult to discover who the legal owner of a company is.

1.3 INSTRUMENT-FREE FIELD SCREENING TECHNOLOGIES

The WHO published the ASSURED criteria as guidelines for diagnostic devices for use in LMICs, and these criteria are equally applicable for field screening technologies: they should be affordable, sensitive, specific, user-friendly, rapid, robust, equipment-free, and deliverable to the end user.[29] In 2019, real-time connectivity, ease of specimen collection, and environmental friendliness were added as additional considerations for device development; the list now goes by the abbreviation REASSURED.[30]

Several reviews of field screening technologies for pharmaceuticals in LMIC settings have been published, with recent examples including Kovacs *et al.,* 2014, [31] Vickers *et al.,* 2018, [32] and Roth, Biggs, and Bempong 2019; [33] and the reader is referred to these for additional connections to the primary literature. This chapter introduces commonly used field screening technologies that are equipment-free. Instrumental methods of pharmaceutical screening and analysis are covered in the following chapters.

1.3.1 WHAT ANALYTICAL METRICS ARE APPROPRIATE FOR FIELD SCREENING TECHNOLOGIES?

The first step in any analytical process is to define the question you are trying to answer. This determines which analytical metrics should be used to evaluate the process. For example, many analytical methods are designed to produce accurate results at low- concentrations of the analyte; this type of high sensitivity is useful for applications such as therapeutic drug monitoring, where blood concentrations of the drug in the ng/ml range may be sought. However, pharmaceutical dosage forms of small-molecule drugs usually contain hundreds of milligrams of the target molecule, so the ability to detect nanogram quantities of the drug is not useful. Low limit of detection is almost never a critical metric for field screening. A more relevant analytical question is whether the technique can resolve a product which contains 86% of the stated API dose, from a product that contains 91% of the stated API dose. For most compendial standards, 86% API content sample fails the content requirement, and 91% API content sample passes the content standard. This difference is extremely important from a regulatory standpoint. There is currently no screening technology available that can identify substandard products reliably in field settings.

1.3.2 PACKAGING INSPECTION

Packaging inspection serves three sometimes overlapping purposes: detection of products that do not meet regulatory requirements (e.g., past date, missing required information such as batch number, or product not registered for sale), detection of products that may be falsified (appearance of label imitates label of an authorized product), and detection of products that may have been manufactured or distributed without due care (incomplete seals on blister packages, boxes falling apart because of bad glue, lack of required patient information leaflets, signs of water damage). In addition to direct visual examination, packaging inspection may involve comparing the look and feel of the packaging materials to an authentic sample, viewing the packaging or pills under light of different colors or under ultraviolet light, as done in the CDC's CD3 device, [34] or more complex instrumental methods.

There is a large industry devoted to packaging security features, ranging from tamper-resistant or tamper-evident seals to bar codes, UV inks, holographic stickers, and micro-embossed features printed on individual tablets. These features can make individual products more difficult to copy, but they do not solve the global problem of SFPs in the market, and all are subject to more or less competent imitation by counterfeiters. Since even pharmacists who regularly handle the packages are not skilled at detecting fake versions of pharmaceutical packaging, fakes can achieve high market share. In the early 2000s, up to 40% of the market for artesunate in some countries was found to consist of inactive counterfeits. Over 15 generations of fake packaging, the counterfeiters gradually improved in the "quality" of their imitation of the security features, until detection of the fakes required a knowledgeable inspector armed with a microscope. [35] In 2010, Sproxil developed a packaging security device that consists of a scratch-off label concealing a code; when that code is texted to a central phone line, the user receives verification of the product on which the label is affixed. Other companies have developed similar security devices. However, this security feature must be incorporated on the product packaging by the manufacturer, and it is vulnerable to imitations. Packaging security features can also be frustrated by placing SFPs into authentic packaging materials (perhaps after selling the authentic pharmaceuticals).

Packaging-based field screening is inexpensive and rapid, and photographs can be used to preserve results of the tests. However, this field screening method is fundamentally limited because it does not assess medicine quality directly, but uses packaging as a proxy for medicine quality. Thus, it is not a reliable way to identify either good or bad-quality products.

1.3.3 Glassware-Based Field Color Tests

Chemical color tests that detect functional groups found in pharmaceutical molecules are the basis for many presumptive tests. These color tests can be conducted in a variety of ways, ranging from quantitative spectrophotometric assays, which require a relatively high-level lab facility with instrumentation and volumetric glassware, to qualitative test tube and spot tests that can be performed in improvised laboratory settings. The analytical literature contains thousands of papers describing quantitative and qualitative chemical color tests, but most of them are "mono-taskers" applied to a few pharmaceuticals. These single-drug color tests are useful in specific cases where deceptive counterfeits of a known product are suspected to be circulating.[36]

The WHO published a series of three monographs listing color tests that are useful in initial screening of hundreds of pharmaceutical active ingredients and dosage forms.[37] These monographs, published between 1986 and 1998, include directions for conducting multiple color tests on each targeted drug, as well as descriptions of reagent preparation and helpful hints about how to conduct the tests in a simple laboratory environment. Many of these color tests were incorporated into the first generation Minilab™.

FIGURE 1.4 WHO monographs on chemical color tests for presumptive testing of pharmaceuticals.

Color tests are particularly useful for detection of pharmaceutical products that do not contain the stated API or contain an alternative API. They can also help to identify unusual excipients. However, qualitative color tests are not able to reliably identify substandard products, and quantitative color tests that can identify substandard products require lab facilities such as volumetric glassware, accurate balances, and UV-Vis spectrophotometers, and are generally not suitable for field use. Many color tests involve hazardous reagents such as concentrated acids, strong oxidizing agents, or noxious gasses, so safety and waste disposal are barriers to implementing these tests in a field situation. The results of field-based tests must be evaluated by eye (subjective interpretation) and recorded manually; there is usually no way to review the record of the test result directly. The prevalence of smart phones offers a solution to this reporting limitation. One major limitation is that color tests are vulnerable to both positive and negative interference from other materials that may be present in the dosage form. Interferences can cause false positive results, false negative results, or colors that are not interpretable. Obtaining information about potential interferents from the packaging materials is usually not possible because they usually do not list excipients, and it is often difficult to find it out from the manufacturers, who may regard it as a trade secret or simply be unresponsive to researchers. The main limitation of color tests is that every test on every drug substance must be performed and interpreted in a different way, requiring a highly trained user to select which test to use, do the test correctly, and interpret the results.

1.3.4 Thin-Layer Chromatography

Chromatography is the science of molecular separations, and is the foundation for analytical methods such as gas chromatography (GC), column chromatography, high-performance liquid chromatography (HPLC), and thin-layer chromatography (TLC). Although there are field-portable GC and HPLC instruments, they are relatively expensive and their main applications outside the lab setting tend to be security-related applications such as detection of explosives, illicit drugs, and chemical warfare agents. Due to its importance in the Minilab™, TLC is the most widely used chromatographic technique for rapid screening of pharmaceuticals in field settings.

Thin-layer chromatography (TLC) uses solid-liquid partition chromatography to separate analyte molecules. The solid component is a thin film of powdered adsorbent, such as silica, alumina, or microcrystalline cellulose, mixed with binders and sometimes fluorescent dyes to help visualize the positions of the analytes. The adsorbent film is supported by a plate of glass, plastic, or aluminum, and usually cut into plates about 2 x 6 cm in size. The liquid component or developing solvent is usually a mixture of polar and non-polar organic solvents, sometimes containing water, acids, or bases to modify the partition behavior of the analytes. The analytes are dissolved in a spotting solvent, typically at mg/mL concentration, and a small spot is deposited near the bottom of the TLC plate. The bottom edge of the TLC plate is placed in the developing solvent inside a jar or tank, so the atmosphere can be saturated with developing solvent, and the liquid rises up the TLC plate by capillary action, bearing the analyte molecules along. The plate is allowed to develop until the solvent front is near the top of the plate (but not at the top). Depending on the solid-liquid partition behavior, the analyte molecules may remain at the spotting line, move partway up the plate, or be carried along at the solvent front. After the developing solvent evaporates, the analyte molecules can be visualized with a variety of methods, some of which also generate chemical color test information. By spotting a sample of the authentic analyte next to the unknown sample, the mobilities and chemical reactivities of the spots can be compared.

TLC is a common technique for chemists, and it is a powerful tool for identification of almost every type of API used in pharmaceutical manufacture. [38] It is relatively easy to develop a TLC method for a new substance and to communicate that technique to other chemists. However, as any organic chemist who has trained students in the lab knows, TLC requires a high level of skill to perform and interpret correctly, particularly for quantitative work. Most TLC plates take about 15 minutes to develop, and the need for consumable supplies such as reference materials, TLC plates, and mixtures of specific organic solvents is a practical barrier for use in LMIC settings. The GPHF Minilab™ addresses some of these barriers.

High-performance TLC (HPTLC) is a relatively recent development that uses ultra-high porosity adsorbent films, carefully controlled sample application, equilibration of the TLC plate with the tank atmosphere, and mechanical insertion of the plate into the developing solvent to achieve excellent chromatographic separations for pharmaceutical dosage forms.[39] The results that can be obtained by HPTLC are in many cases comparable with those achieved with high-performance liquid chromatography. However, these HPTLC systems are benchtop instruments like HPLC systems. Like HPLC, they are expensive and delicate, and are not currently suitable for field screening.

1.3.5 MERCK-GPHF MINILAB™

A GPHF Minilab™ is a "suitcase lab" that contains supplies and reference materials for testing over 100 essential drugs (Figure 1.5). The screening activities are based on the WHO (1999) publication, "Guidelines to Combat and Test Counterfeit Medicines."[40] The recommended screening activities have undergone two major iterations. The initial minilab screening process started with physical inspection of the packaging and dosage units, and weighing the tablets or capsule contents. Next, several chemical color tests were carried out in miniature glassware. The results could either be compared to photographs, or the user could run the chemical color tests in parallel on reference substances provided in the kit. Tablets were evaluated with a simplified disintegration test as an alternative to complex and expensive compendial dissolution testing. In a disintegration test, the tablet or capsule is placed in a quantity of water for a set time. If the tablet does not disintegrate, it is

FIGURE 1.5   Marya Lieberman with Richard Jähnke, inventor of the minilab, showing one of the two suitcases packed with reference materials and lab supplies in 2018. Photo: Marya Lieberman.

a possible sign that the tablet will not disintegrate in the patient's stomach, which could prevent the drug from being absorbed properly. Finally, quantitative thin-layer chromatography is performed using reference materials provided in the form of tablets. These reference tablets are dissolved in a known volume of solvent, providing solutions of known concentration to calibrate the TLC results. In the current minilab process, the chemical color tests have been de-emphasized, and TLC is the main tool recommended for identification and quantification of 107 active pharmaceutical ingredients.[41] Quantification is performed by comparing the intensity of the sample's TLC spot with the intensity of the TLC spot from a reference at 100% and a reference at 80% of the expected API concentration (Figure 1.6). If the sample spot intensity is less than the 80% reference, the sample is reported as a suspected substandard product. This protocol is not expected to detect products that are only slightly substandard, e.g., containing between 80% and 90% of the expected API content.

The GPHF Minilab™ ships as two plastic cases weighing 50 kg and costs about $7,400 for materials and reference substances suitable for 1,000 tests. Some of the supplies, like the UV lamp which helps to read the spot locations, are electrically powered (line current or battery powered) An introductory training class for users lasting a week can be supplemented with additional training to sharpen the user's TLC interpretation skills.

The evidence base shows both strengths and weaknesses of the minilab. A study involving 13 countries and over 1,900 samples had a low estimated cost per sample of €25 for minilab testing.[42] However, several studies have reported that even

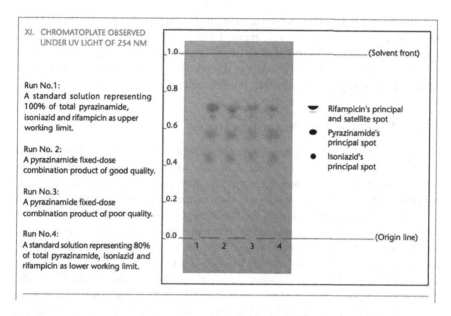

FIGURE 1.6 Minlab manual page showing directions for evaluation of pyrazinamide in a fixed-dose pyrazinamide, isoniazid, rifampicin tablet. The TLC method can analyze all three of these drugs, but obtaining and interpreting the results requires a skilled operator.

semi-quantitative analysis with the minilab is difficult for staff to do accurately. In a WHO survey of antimalarial quality across six countries in Africa, the minilab detected only one in three of the products classified as "extremely non-compliant". [43] In another study, 451 samples were analyzed by the minilab; of the 14 samples whose API content was below 50%, six were detected as substandard.[44] The test accuracy is strongly dependent on the skill and level of training of the staff who administer the tests, with a minimum of one week recommended for non-quantitative analysis. In a study in Tanzania, after a five-day training session, then proficiency tests with 0%, 40%, and 100% API, 28/28 of 0% samples were found, but only 3/28 of the 40% samples were found.[45] After retraining with calibrated capillary tubes, standards at 100% and 50%, and explicit practice in interpretation, 19/27 of the 40% samples were found. Recently, a phone app for interpretation of the TLC spot locations and intensities has been reported; this innovation has the potential to improve accuracy of reading the TLC results and could also be used to archive the raw data.[46]

GPHF Minilabs™ have the strongest record of implementation and utility in LMIC settings of any pharmaceutical screening technology.[47] More than 900 minilabs have been dispersed to 99 countries, including many LMIC sites. Minilab testing has resulted in dozens of WHO Medical Product Alerts. The analytical strength of the minilab system is its generalizability to multiple pharmaceutical types, including those that do not generate strong chemical color tests due to their lack of targetable functional groups. New drugs can be added to the minilab by developing and testing a new TLC method and providing the reference materials. The main limitation of the minilab as a field screening technique is that the whole process requires a high level of skill in the operator, particularly for quantitative work, and the equipment is movable, but not really portable.

1.3.6 DISINTEGRATION/DISSOLUTION TESTING

In order for oral medicines to achieve the desired effect, they must dissolve in the user's gastrointestinal (GI) tract – usually in the stomach. Tablets that are formulated, pressed, or coated incorrectly may pass completely through the patient's system intact, in which case they cannot deliver the benefits expected from the medication even though the tablet might contain the correct dosage of API. Monograph methods include detailed procedures for dissolution testing, which uses a special instrument to simulate the acidity and churning action of the stomach. The rate of release of the API is measured to determine whether the formulation is likely to dissolve properly inside a patient's body. Currently, dissolution testing is not "fieldable" as a screening test method. Disintegration, as used in the minilab, can be performed as a field test; a simplified version is done by placing the pill to be tested in a cup of water for a specific time, then weighing any undissolved materials. A prototype device, the PharmaChk, is under development by the Zaman group and has promise for use in regulatory labs, but is not yet easily usable in field settings.[48]

1.3.7 LATERAL FLOW IMMUNOASSAY STRIPS

Antibody based immunoassays are commonly used as medical diagnostics; COVID-19 test strips, malaria RDTs, and pregnancy test strips are familiar examples. While these test strips have great potential for evaluation of biologics and very low-concentration medications, their mode of action requires different antibodies for detection of different targets, which presents logistical challenges for field analysis of multiple types of medicines. Polyclonal antibodies (from the serum of challenged animals) can be used for test development, but unless monoclonal antibodies are used for manufacturing, reproducibility of sensitivity and specificity for the target is difficult to achieve.[49] Since the strips are able to detect very low quantities (typically ng/ml) of target, they are often used for environmental or food quality testing, for example detection of antibiotic residues in meat or milk.[50]

One example where lateral flow immunoassay testing has had considerable impact on a drug-quality question is not in an LMIC setting, but in the US, for use in detection of illicit drugs that contain the superpotent opioid fentanyl.[51] It is not surprising that this use case has emerged. The illicit drug trade in the US is completely unregulated and is served by manufacturers who are by definition criminal – quality assurance (QA) and quality control (QC) are not a high priority. The fentanyl molecule is too small to use in a sandwich immunoassay, so fentanyl test strips use a competitive immunoassay strategy; this strategy can be applied to other types of small molecule drugs.[52] Problems with lateral flow test strips for drug screening applications include false positives caused by off target binding (particularly at the relatively high concentrations obtained by field sample preparation), [53] inability to quantify the target, and variations in results caused by poor QA/QC on the strips themselves.

1.3.8 PAPER MICROFLUIDICS: THE PAPER ANALYTICAL DEVICE (PAD)

Some field screening devices for pharmaceuticals use hydrophilic paper substrates patterned with hydrophobic barriers to carry out chemical tests.[54] These devices may be called paper microfluidic devices, paper analytical devices, or micro total analytical systems (μTAS). [55,56] They are typically inexpensive single-use devices which have many advantages for use in low-resource settings.[57] Most paper analytical devices for pharmaceutical screening are laboratory devices that may have undergone some field testing, but are not commercially available. One type of paper analytical device (PAD) designed for pharmaceutical screening [58] is made in my lab at the University of Notre Dame. [59] I'll draw on this example to illustrate the technology development process.

1.3.8.1 How the PAD Works

The PAD contains 12 lanes of chromatography paper that are separated by wax-printed hydrophobic barriers. Each lane contains different chemical reagents that are stored in dry form in the paper. Users swipe a pill across the paper to deposit a small amount of the pill material in each of the 12 lanes, then dip the bottom edge of the PAD into water to redissolve the stored reagents and move them up the lanes. (Figure 1.7) These reagents react with different chemicals in the pill, creating a distinctive color bar code. Users can read the bar code by eye, and we have developed software for automated analysis of the paper tests based on cell phone images.[60] The software makes it easy to evaluate the results and enables better record keeping and tracking of low quality medications across large regions.

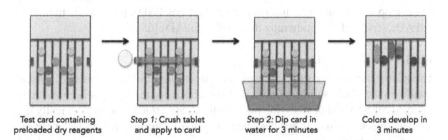

| Test card containing preloaded dry reagents | Step 1: Crush tablet and apply to card | Step 2: Dip card in water for 3 minutes | Colors develop in 3 minutes |

FIGURE 1.7   How the Paper Analytical Device (PAD) is used to test pharmaceuctical dosage forms.

1.3.8.2   Test Development

There is a rich literature on the adaptation of laboratory analysis methods to paper microfluidic devices. [61,62] In general, control of flow, sample preparation and addition, reagent stability, and signal transduction are all critical features. From an analytical standpoint, the reproducibility of results, robustness of the process over time and experimental conditions, and tolerance for interferences are critical metrics.

Translation of glassware based tests to the PAD format typically occurs in several stages. The first step usually involves identifying a published method that generates a color result for the target pharmaceutical or for a functional group found in the target pharmaceutical. The next step is to adapt the method to accommodate the constraints of the wax-printed paper card. For example, many analytical procedures call for non-aqueous solvents, which would dissolve the wax barriers on the paper card, and common reagents like concentrated sulfuric acid or cerium(IV) cannot be stored on paper because they react with the cellulose. In some cases, milder reagents can be substituted, such as p-toluene sulfonic acid for hydrochloric or sulfuric acid. Potential reagents are first tested as spot tests on paper substrates to identify a reagent set and positive and negative control substances, then the reagent(s) are deposited in paper lanes and the optimal location of the reagent(s) within each lane are determined. The lane is then tested with a dosage form of the target pharmaceutical and with diluted dosage forms to determine the approximate limit of detection for the target analyte. A wide variety of pharmaceutical APIs is screened on the new lane to determine possible cross-reactions and interferences. It is common for one lane to yield multiple color reactions with different analytes, often in different parts of the lane, and of course, a single pharmaceutical can trigger color reactions in multiple different lanes.

The current PAD reagent set can detect over 60 different drugs, such as amoxicillin, ampicillin, ciprofloxacin, isoniazid, and metformin, as well as fillers used in falsified medicines. In some cases, small functional group differences between chemically similar drugs produce very different color bar codes (Figure 1.8). In other cases, related drugs produce indistinguishable bar codes, and if a pharmaceutical does not contain functional groups that react with the set of reagents present on the PAD, the pharmaceutical can't be evaluated. Although we have tried to make the PAD as general as possible, additional reagent sets had to be developed for use with antimalarial drugs (Laos PAD), [63] illicit street drugs (idPAD), [64] and chemotherapy drugs (chemoPAD).[65]

1.3.8.3   Making PADs

There are many methods for fabrication of paper microfluidic devices but most are used to make small numbers of devices for laboratory testing.[66] For production on scales of >10,000 devices, the main methods are screen-printing, typically used to make electrochemical devices such as glucose sensors; roll-to-roll processes used for production of paper dipstick tests or nitrocellulose/paper lateral flow assays, and wax printing, used to make paper analytical devices for diagnostic assays [67] and pharmaceutical testing. [68] We have made over 50,000 test cards

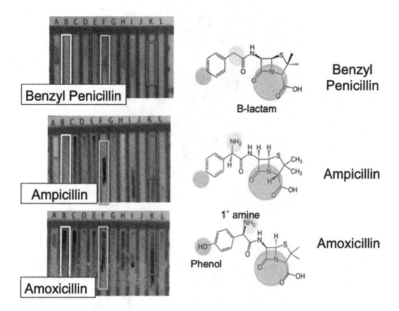

FIGURE 1.8 Active pharmaceutical ingredients with very similar chemical structures can give very different color bar codes on the PAD. Lane F contains basic copper(II), which gives a dark green color in the presence of the beta lactam group present in all three antibiotics. Lane B contains ninhydrin, which reacts with the primary amine groups in ampicillin and amoxicillin to form a yellow or green color, respectively. Lane K contains a test for the phenol functional group, which generates the dark red color for amoxicillin.

in the past five years using wax printing to form the hydrophobic lane barriers and hand stamping to deposit the reagents in the lanes (Figure 1.9). The card print files include individual serial numbers and QR codes; the serial numbers can be linked to a project, which simplifies tracking of results. The cards fit eight to a page and are printed on Ahlstrom 319 or IW Tremont DBS1 cellulose papers. The print files are downloaded, then the cards are printed on the front with a laser printer, printed with wax ink on a Xerox ColorQube wax printer, and baked at 100°C for about six minutes to enable the wax to reflow and create hydrophobic lane barriers. Next the cards are stamped with reagents. The pitch of the lanes is 4.5 mm laterally, but the pitch of the 96 well plate that holds the reagent solutions is 9 mm. Thus, it takes about 60 seconds to get all of the reagents onto the card because each card must be stamped twice. The cost of materials and labor is $0.47 per card.

1.3.8.4 Data Analytics for the PAD

The initial process for the PAD required users to read the results by visually comparing the color bar code to either standard images or a "fringe diagram" showing the range of acceptable outcomes in each color-generating lane. This

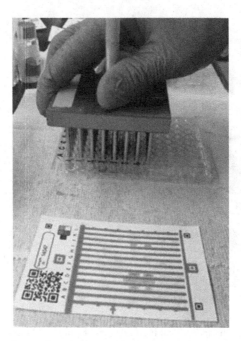

FIGURE 1.9 A hand stamping tool is used to deposit reagents on a PAD. The grey lines are hydrophobic barriers formed by wax printing.

method worked well for "expert" users who were familiar with the normal range of colors. In fact the pattern-matching ability of the human visual system enabled some users to identify useful new color signatures. In the case of one antibiotic, pharmacists screening PADs in Kenya noticed a new color in one lane of thermally degraded samples. We found that when the antibiotic degraded, it raised the solution pH in an autocatalytic manner, and the basic pH triggered the new color in one of the lanes. [69] However, for every group of nonexperts with whom we worked, reading the PAD was the hardest part of using it, and the task became more challenging the more types of APIs we asked our users to identify. Since the PAD is chemically capable of distinguishing over 60 types of APIs, we knew that visual recognition would not be scalable.

We were interested in automating the classification of the API types and using the color tests to extract some quantitative, or at least semi-quantitative, information about the API targets. Fortunately there are many good data analysis methods at hand for this task.[70,71] The simplest method is to make a calibration curve. At very low concentrations of the target, spot tests give negative outcomes. At the limit of detection, the spot test color can be detected by eye, and typically there is a range of concentrations over which the color intensity increases as the target concentration increases. At high concentrations of the target, the color saturates. Linear or nonlinear calibration curves can be prepared by extracting the grayscale pixel intensities in a region of interest, typically with an image analysis program like Image J. The pixel intensities in the red, green, and blue channels can be extracted from multiple regions of interest and analyzed with more sophisticated linear methods such as principal component analysis or other machine learning approaches. This data analysis approach was applied to the PAD for detection of substandard antimalarial formulations; we found that even in a controlled lab setting, we could not be sure of detecting products with just 50% of the stated API content. The problem is not the data analysis or the color reaction itself, but the amount of solid that is applied to each lane of the PAD. In recent years, neural networks have become easier to apply to image recognition problems, and this approach has shown promise for PAD analysis.[72]

1.4 AN IMPLEMENTATION CASE STUDY: PAPER ANALYTICAL DEVICE (PAD)

Implementation issues are discussed here in the case of one novel technology, the paper analytical device (PAD). This technology went through a process of laboratory development, internal lab validation, and field testing, as described in the previous section. However, implementation requires making the technology fit the real-world context of regulations and commercial activities. For another example of device development, visit the tutorial review on the PharmaChk device. [73]

Intellectual property (patents or trade secrets) is of potential importance in two areas: first, preventing others from competing, either to profit from a technology or to ensure that bad-quality copies of the technology are kept out of a market; and, second, ensuring that the license holder retains freedom to operate if the technology space includes other products protected by IP. Public disclosure of a technology (publication or presentation at a conference) will harm patentability; a provisional US application can be filed quickly to protect the potential IP for 12 months. Evaluating patentability is difficult and consultation with a patent lawyer is recommended. Most academic institutions have limited resources for pursuing patents and hence will take a close look at the potential profitability of the proposed technology before doing more than filing a one-year provisional US application and possibly a PCT provisional international application. A US patent application for the PAD was filed in 2012 with the generous pro bono support of Winstrom and Strawn, a major US patent firm, and the final patent was issued in 2016.[74]

Regulatory approval of specific field screening devices is a common sticking point for implementation of field screening tests used outside the setting of a regulatory drug lab. Drug regulators operate in a lab environment that is heavily regulated, and they often request information about the regulatory status of a screening device as an assurance that it will produce credible results. However, there is no global regulatory authority for pharmaceutical field screening tests. The WHO issues the IP, which contains detailed metrics that can be used to evaluate laboratory analyses, but it does not provide guidelines to evaluate field screening methods. Within each country, the regulatory situation and applicable regulatory bodies are different. In the US, the Food and Drug Administration (FDA) is responsible for overseeing medicine quality and medical devices. When we applied for an Investigational Device Exemption to cover a field test of the idPAD device in Chicago, we learned that FDA defines "medical devices" as anything intended to diagnose or cure diseases or to affect the structure and function of the body.[75] Their determination was that the idPAD did not fit this definition, and it is likely this reasoning would apply to many of the other field test methods described in this present manuscript.

Regardless of the regulatory status of the device itself, articulation of any field screening project or program with relevant regulatory authorities is necessary in order to achieve impact. Unless the screening program is being conducted by the drug regulator itself, the regulator needs to be in the loop to ensure that suspected bad-quality products receive proper confirmatory testing and enforcement activities are conducted when appropriate. It is critical to build relationships with regulatory agencies and to obtain funding to support exploratory or pilot tests of the technology by drug inspectors or lab technicians, who already have their hands full with their day to day tasks. Transnational organizations such as Interpol, WHO, and the USP's PQM+ program are generally focused on threat communication and capacity development, and Interpol coordinates joint enforcement activities with national drug regulatory agencies. All of these transnational organizations work closely with national drug regulators and may be able to provide venues for pilot tests.[76,77]

Manufacturability is a key consideration even during laboratory development of field screening devices. If a device is easy to make at scale, it is much easier to implement it in field testing and early stage commercialization activities. For example, novel immunoassays are relatively easy to manufacture. There are small-scale reagent deposition, assembly, and cutting tools that can make several thousand strips for field testing and can be scaled up to faster machinery to make hundreds of thousands of strips to distribute as a product. In contrast, there is no established infrastructure for manufacture of PADs, so we knew early on that we would need to make hundreds or even thousands of PADs in a research lab setting. The PAD was therefore designed to facilitate pilot-scale manufacture.

(1) A commercial wax printer was used to print the hydrophobic barriers; eight PADs fit on each sheet of Ahlstrom 319 paper with no waste paper.

(2) The physical dimensions of the lanes were designed to be compatible with the 9mm pitch of a 96 well plate, which meant that the reagents stored on each lane of the PAD could be deposited using either a manual multi-spoke deposition tool, or a 96 well plate spotting robot.

(3) Compatibility with digital image analysis was baked into the cards via inclusion of fiducial marks and generation of unique serial numbers and QR codes for each card, and

(4) QA/QC procedures for preparation and testing of reagents and finished PADs were developed to ensure consistency in PAD performance. Between 2015 and 2022, more than 55,000 PADs were manufactured in-house with these procedures.

User training was initially a multi-hour process because users had to read the color bar codes by eye. In 2018, 20 drug inspectors at the Tanzanian Food and Drug Authority (TFDA) received a three-hour PAD training session. Over the next two months, they were collectively sent 3,000 PADs and 3,000 samples of five medications (amoxicillin, benzyl penicillin, penicillin-procaine, paracetamol, and quinine). Half of the samples were high quality formulations and half were formulated with the wrong API or with no API. The analysts used PADs to distinguish between the good quality and bad quality samples and they uploaded cell phone images of the PADs to a Dropbox folder. Analysts felt confident in their ability to use and read the PADs. They appreciated the rapid analysis time and portability for field analysis but wished the PADs could identify substandard products. Some inspectors reported difficulty in interpreting some colors. Inspectors agreed that the PAD could be useful for TFDA to use during post-market surveillance or for screening at ports of entry.

Automation of the readout of the PAD via machine learning approaches has greatly reduced the complexity of user training and enhanced the usability of the PAD for data collection and sharing. TFDA analysts had an overall accuracy of 92.6% in distinguishing good and bad formulations, while an "expert" PAD reader evaluating the blinded PAD images had an accuracy of 97%. The uploaded PAD images were used to train a neural network using the "Caffe Net" architecture;[78] the neural net was able to correctly classify samples as "consistent with label" or "suspicious" with an accuracy of 99.17%.

The neural net was incorporated into mobile apps for Android and iOS operating systems in 2021, after the LOMWRU and TRP field tests (see Section 1.5). The mobile app recognizes the PAD using the cell phone or tablet camera, captures the image, and prompts the user to enter additional metadata such as the brand name and lot number. The app reads the test result and reports immediately to the user. The raw data is stored on the phone until it finds a good Wi-Fi connection, at which point the data and test results are uploaded to a back end database. These mobile apps are designed to meet the principles of the Open Science movement, allowing access to the test results to be easily shared outside the proprietary systems that generate the results.

1.5   WHO TESTS THE TESTS?

The evidence for effectiveness of a field screening test for pharmaceuticals can be demonstrated in different ways. The most basic way is through an internal laboratory validation, which usually uses samples that are mocked up from pure API mixed with common pharmaceutical excipients such as microcrystalline cellulose, starch, or lactose. The API content and excipients should correspond to examples of good quality, substandard, and totally fake formulations. These samples should be blinded so the people who do the test and interpret the test results do not know what category the sample belongs to. Real dosage forms collected in the country of interest can also be analyzed, although their purity cannot be assumed – there are multiple cases where clinical trials [79] or analytical training sets [80] were corrupted by inadvertent inclusion of an authentic product that did not meet quality standards. Field-collected samples used to evaluate field screening tests should always be assayed with methods known to be accurate, such as a compendial HPLC method.[81]

A critical step in the validation process is an external validation study, where performance tests are repeated by naive users or by users in another laboratory. This type of study is very helpful for identifying operational or training requirements that the test developers in their home laboratory may not have realized are important, so doing it in several phases is a good experimental design. Two organizations have performed external validations for a wide range of field screening technologies for detection of SFPs. Both included PADs, and it is of interest to compare their methodology and conclusions.

1.5.1   LOMWRU Multiphase Review

In 2018, Caillet *et al.,* reported on a large multiphase review of the accuracy, utility, usability, and cost-effectiveness of various portable devices intended to identify SFPs, particularly anti-malarial and antibacterial drugs used in SE Asia.[82] The project was conducted by the Lao-Oxford-Mahosot Hospital-Wellcome Trust Research Unit (LOMWRU), funded by the Asian Development Bank, and was conducted in parallel with the early stages of the USP Technology Review Program. It is a rare example of an interlaboratory comparison of device performance for field screening technologies.

In the first phase of the project, a literature review [83] identified 41 devices for screening of pharmaceuticals, 27 of which were portable spectrophotometers – 14 of the devices were selected for the next phase. The most common reasons that devices were excluded from the next phase were that they were mono-taskers (the median number of APIs described in published reports was two) or, in the case of vibrational spectrophotometers, required the operator to use or generate "complex API-specific calibration models." The LOMWRU team requested PADs for testing a specific list of antibiotics and antimalarials, and since these drugs were not used as much by our partners in Africa, we developed a custom PAD using samples of dosage forms of the targeted drugs. Training materials and hundreds of the new PADs were delivered to the lab-based evaluators at Georgia Tech, and they trained the field testers in Laos.

In the laboratory phase, which was conducted at Georgia Tech, a performance evaluation of 12 devices was carried out to determine their analytical metrics and assess their suitability for the field phase.[84] Samples included field-collected pharmaceutical dosage forms and lab-made SF samples. The devices tested included "three near infrared spectrometers (MicroPHAZIR RX, NIR-SG1, Neospectra 2.5), two Raman spectrometers (Progeny, TruScan RM), one mid-infrared spectrometer (4500a), one disposable colorimetric assay (Paper Analytical Devices, PAD), one disposable immunoassay (Rapid Diagnostic Test, RDT), one portable liquid chromatograph (C-Vue), one microfluidic system (PharmaChk), one mass spectrometer (QDa), and one thin layer chromatography kit (GPHF-Minilab™)." All of these devices performed well for detection of samples with authentic compositions versus no API or the wrong API. However, several were assessed as too difficult to use in the field phase due to the high levels of training required, the need for special or expensive consumables, anticipated export problems, or problems in sourcing the necessary numbers of devices.

For the field phase of testing, a mock pharmacy was constructed on the grounds of Mahosot University in Lao PDR, and 16 drug inspectors conducted utility and usability tests using six devices.[85] The devices included four handheld spectrometers (two near infrared: MicroPHAZIR RX, NIR-S-G1 and two Raman: Progeny, Truscan RM); one portable mid-infrared spectrometer (4,500a), and single-use paper analytical devices (PAD). Additional testing was conducted in office settings with the devices and in a lab with the GPHF-Minilab. In the pharmacy setting, the drug inspectors had difficulty in reading the PAD results, misclassifying about a quarter of the samples. However, they liked the PAD's lack of reliance on electricity or sophisticated instrumentation, and felt that manufacturers might use it to test raw materials. The PAD and minilab, which both require sample preparation and chemical operations, took significantly longer to test a sample (10 minutes and 34 minutes, respectively) than the spectrophotometers (1.5–5 minutes).

The cost-effectiveness of the six devices tested in the field phase was estimated for scenarios that assumed either high or low levels of SFPs for a particular set of medicines. All six devices were predicted to be cost-effective for the high level scenario, and four (MicroPHAZIR RX, 4500a FTIR, NIR-S-G1, and PADs) were cost-effective for the low-level scenario.[86] These cost estimates depend on assumptions such as how many samples are tested each year; the more samples are tested, the more cost-effective the portable spectrophotometers (which have large capital costs) become. The final phase of the study was a stakeholder meeting that summarized lessons learned and evidence gaps.[87]

1.5.2 USP Technology Review Program

A US Pharmacopoeia monograph describing the procedures for evaluating screening technologies was issued in 2020, with input from the Expert Panel on Review of Surveillance and Screening Technologies for the Quality Assurance of Medicine.[88] It lays out six target applications, each with suggested analytical performance characteristics that should be assessed, and systematically lists important questions related to the practicality and costs of the technology. The application areas are: verification of packaging, labeling, origin, and appearance; major component ID; impurity ID; major component quantification; impurity quantification, and performance characteristics such as dissolution behavior. The monograph suggests that the device performance should be assessed in both lab and field settings for any device intended for field screening. The monograph does not provide testing methods or metrics for the suggested analytical performance characteristics; these experimental design decisions are left to the external laboratories that are conducting the lab and field evaluations.

Concurrently with development of USP General Chapter <1850>, the USP Technology Review Program (TRP) was set up to evaluate promising screening technologies for use in LMIC settings. From 2017 until 2020, this program issued evaluation reports for six screening technologies: the PAD, Target-ID (a portable Fourier-transform infrared spectrophotometer), Speedy Breedy (a device for testing sterility of liquids), the CBEx Handheld Raman Spectrometer, the ASD QualitySpec near-IR spectrophotometer (Trek), and the Global Pharma Health Fund (GPHF)-Minilab™. Each technology was tested in lab and field settings in LMICs, and the rest results are published at the USP TRP website.

When the TRP selected the PAD for evaluation, the first step in the process was to **define the target application** of the technology, such as the types of pharmaceuticals it could test and whether it was intended to detect falsified or substandard products. Without a clear use case, it is not possible to fairly evaluate the performance of any screening technology. Application II, Major component ID, was selected. Next, the **specifications, cost elements,** and **outputs** of the technology are systematically gathered, based mostly on manufacturer information; in the case of the PAD, USP requested this information from our lab. These considerations help potential implementers compare different technologies, and allow the selection of analytical performance characteristics and the design of experiments to evaluate those performance characteristics. The **performance evaluation** of the PAD was conducted in the USP laboratory in Ghana. At the time (summer of 2020), USP was particularly interested in the performance of the PAD for analysis of chloroquine, hydroxychloroquine, and azithromycin, which had been touted as miracle cures for COVID-19. Although these had not been among the drugs for which the PAD was optimized, when we ran test samples, we found that the PAD gave usable color bar codes for all three. Amoxicillin, ciprofloxacin, ceftriaxone, and doxycycline formulations were also evaluated. Three different brands of these seven finished pharmaceutical products were tested, along with the seven pure APIs, 14 lab-made falsified samples, and 14 lab-made substandard drug formulations. Each sample was analyzed in triplicate by three different scientists. Finally, a **field evaluation** was carried out in the Ugandan national drug authority's lab and in various non-lab settings, to evaluate the robustness of the technology and how long it took for users to become proficient with the technology. Eight staff members with varying levels of experience were trained and the PADs were used in a retail pharmacy, a national general hospital, and a central medical warehouse. At this stage, the phone app/PADreader was not yet available, so the users had to compare the PAD color barcodes to photographs of standard images provided for each API.

Results from the review: [89] "The PAD was able to identify the active ingredients tested in all the brands of pharmaceutical FDFs and in their respective pure raw materials as they showed the appropriate colors in their respective lanes for all three of the scientists. The PAD was able to detect fillers such as corn starch in some tablet dosage forms, namely: azithromycin, chloroquine, and all three brands of hydroxychloroquine. Corn starch was easily seen in lane J as a black color and this agreed with the products information leaflets provided by the drug manufacturers. All falsified formulations tested were correctly identified by the PAD as results from all the scientists were reproducible and comparable." The TRP and LOMWRU participants had similar comments about the difficulty of reading the PADs "by eye" and its unsuitability for identification of substandard products.

While we continue to develop new ways to generate color test results on the PAD, the main push for implementation now is interfacing the physical PAD card with electronic data collection and data management tools. The results of the TRP and LOMWRU studies, along with the earlier comments from the Tanzanian FDA study, inspired us to refine automated image analysis algorithms [90] for reading the PAD, which we accomplished in 2020–2021. The optimized image analysis software was incorporated into a mobile app that can classify various APIs based on the color barcode on the PAD. Versions of the app are available for Android [91] and iOS phones and tablets.[92] The mobile app allows the user to capture sample metadata, such as the brand and batch numbers, and we are currently integrating it with a third party sample tracking and laboratory information management system. [93] Management of the results from field analysis is a serious "missing stair" in field screening methodology, because the devices often have no means of recording sample metadata, archiving raw test data, or sharing results with laboratory information management systems. Addressing these data management challenges is the next step in development of the PAD system.

1.6 CONCLUSION

Pharmaceutical field screening is an important tool for maintaining the quality of medicines in the rapidly changing and fragmented markets typical of LMIC settings. Even though most field screening methods are bad at detecting trace contamination or substandard products, they are still effective in catching products that contain the wrong API, or no API at all, and these products represent an ongoing threat to the effectiveness of medical care in LMICs.

The evaluation of medicine quality requires a continuum of analysis, from community pharmacies and clinics, to regional regulatory offices, to central drug analysis labs and forensic analysis facilities. This continuum can support different types of field screening methods, ranging from simple methods of packaging analysis or packaging security features, to non-instrumental screening tools like the PAD, to portable vibrational spectrophotometers and suitcase labs like the minilab, to standard analytical instrumentation that requires laboratory infrastructure and trained operators. The following chapters in this monograph provide an introduction to instrument-based methods of field screening (IR/Raman/NIR, UV/Vis), compendial analysis (HPLC), and more advanced instrumental methods (GC-MS, LC-MS, NMR).

REFERENCES

1   World Health Organization. (2017). *WHO global surveillance and monitoring system for substandard and falsified medical products*. World Health Organization. apps.who.int/iris/handle/10665/326708. License: CC BY-NC-SA 3.0 IGO

2   Dioscorides Pedanius [ca. 70]. Osbaldeston, T. A. (ed., 2000). *De materia medica: Being an herbal with many other medicinal matters.* Written in Greek in the first century of the common era. Vol. 2. Johannesburg: Ibidis. ISBN 0-620-23435-0.

3   Broughton, J. On a false cinchona bark of India. *American Journal of Pharmacy* 1868; July, 350–355.

4   World Health Organization. (2017). *A study on the public health and socioeconomic impact of substandard and falsified medical products*. World Health Organization. apps.who.int/iris/handle/10665/331690. License: CC BY-NC-SA 3.0 IGO

5   Hamill, H., David-Barrett, E., Mwanga, J. R., Mshana, G., & Hampshire, K. (2021, May). Monitoring, reporting and regulating medicine quality: tensions between theory and practice in Tanzania. *BMJ Global Health*, 6(3):e003043. doi: 10.1136/bmjgh-2020-003043. PMID: 34049934; PMCID: PMC8166622.

6   Ozawa, S, Evans, D. R., Bessias, S., Haynie, D. G., Yemeke, T. T., Laing, S. K., & Herrington, J. E. (2018). Prevalence and estimated economic burden of substandard and falsified medicines in low- and middle-income countries: a systematic review and meta-analysis. *JAMA Network Open.* Aug 3;1(4):e181662. doi: 10.1001/jamanetworkopen.2018.1662. PMID: 30646106; PMCID: PMC6324280.

7   Mackey, T. K. & Liang, B. A. (2013). Improving global health governance to combat counterfeit medicines: a proposal for a UNODC-WHO-Interpol trilateral mechanism. BMC Medicine. Oct 31;11:233. doi: 10.1186/1741-7015-11-233. PMID: 24228892; PMCID: PMC4225602.

8   Hamilton, W. L., Doyle, C, Halliwell-Ewen, M., & Lambert, G. (2016, December). Public health interventions to protect against falsified medicines: a systematic review of international, national and local policies. *Health Policy and Planning.* 31(10), 1448–1466. doi: 10.1093/heapol/czw062. Epub 2016 Jun 16. PMID: 27311827.

9   Fernandez, F. M., Hostetler, D., Powell, K., Kaur, H., Green, M. D., Mildenhall, D. C., & Newton, P. N. (2011). Poor quality drugs: grand challenges in high throughput detection, countrywide sampling, and forensics in developing countries. *Analyst.* 136(15), 3073–3082. doi: 10.1039/c0an00627k.

10   US Pharmacopeia-National Formulary, Ampicillin Capsules, USP42-NF37 – 316, GUID-47D1A183-2247-4F49-AE1D-4700817783C7_2_en-US. doi.org/10.31003/USPNF_M4440_02_01.

11   Roth, L., Adler, M. A., Jain, T., & Bempong, D.K. (2018). Monographs for medicines on WHO's model list of essential medicines. *Bulletin of the World Health Organization.* 96, 378–385. doi: 10.2471/BLT.17.205807.

12   Roth, L., Nalim, A., Turesson, B., & Krech, L. (2018). Global landscape assessment of screening technologies for medicine quality assurance: stakeholder perceptions and practices from ten countries. *Global Health.* 14, 43. doi.org/10.1186/s12992-018-0360-y

13   Newton, P. N., Lee, S. J., Goodman, C., Fernández, F. M., Yeung, S., Phanouvong, S., Harparkash, K., et al. (2009). Guidelines for field surveys of the quality of medicines: a proposal. *PLoS Med.* 6(3): e1000052. doi:10.1371/journal.pmed.1000052

14   WHO Expert Committee on Specifications for Pharmaceutical Preparations, WHO Technical Report Series No. 996, 2016, Annex 7: Guidelines on the conduct of surveys of the quality of medicines.

15   Fernandez, F. M., Hostetler, D., Powell, K., Kaur, H., Green, M. D., Mildenhall, D. C., Newton PN. (2011). Poor quality drugs: grand challenges in high throughput detection, countrywide sampling, and forensics in developing countries. *Analyst.* Aug 7;136(15):3073–3082. doi: 10.1039/c0an00627k. Epub 2010 Nov 25. PMID: 21107455; PMCID: PMC3427017.

16   Roth, L., Nalim, A., Turesson, B., & Krech, L. (2018). Global landscape assessment of screening technologies for medicine quality assurance: stakeholder perceptions and practices from ten countries. *Global Health.* April 25; 14, 43. doi.org/10.1186/ s12992-018-0360-y

17   WHO, 22nd Model List of Essential Medicines (electronic format with recommendations for 591 medicines and 124 therapeutic equivalents). Accessible at: https://list.essentialmeds.org/ (accessed 26 Nov 2022)

18   The Global Fund, WHO Alert on Counterfeit Coartem, www.theglobalfund.org/ en/oig/updates/2013-11-08-who-alert-on-counterfeit-coartem/ (accessed 26 Nov 2022); see also: WHO Medical Product Alert N°8/2021: Falsified Combiart at www.who.int/teams/regulation-prequalification/incidents-and-SF/full-list-of-who-medical-product-alerts (accessed 26 Nov 2022)

19   Renschler, J. P., Walters, K. M., Newton, P. N., & Laxminarayan, R. (2015). Estimated under-five deaths associated with poor-quality antimalarials in sub-Saharan Africa. *American Journal of Tropical Medicine and Hygiene.* Jun;92(6 Suppl):119–126. doi: 10.4269/ajtmh.14-0725. Epub 2015 Apr 20. PMID: 25897068; PMCID: PMC4455082

20   FDA, Information on Heparin, accessible at www.fda.gov/drugs/postmarket-drug-safety-information-patients-and-providers/information-heparin (accessed 26 Nov 2022).

21   Nishtar, S. (2012). Pakistan's deadly cocktail of substandard drugs. *Lancet,* 379, 1084–1085.

22   Chaudhiri, S. (2013). Ranbaxy to pay $500 million in adulterated drugs case. *The Wall Street Journal.* Retrieved from: www.wsj.com/articles/SB10001424127887 323716304578481182961557130.

23   Dégardin, K., Guillemain, A., Klespe, P., Hindelang, F., Zurbach, R., & Roggo Y. (2018). Packaging analysis of counterfeit medicines. *Forensic Science International.* Oct, 291, 144–157. doi: 10.1016/j.forsciint.2018.08.023. Epub 2018 Aug 29. PMID: 30205292.

24   Schiavetti, B., Wynendaele, E., Melotte, V., Van der Elst, J., De Spiegeleer, B., & Ravinetto, R. (2020). A simplified checklist for the visual inspection of finished pharmaceutical products: a way to empower frontline health workers in the fight against poor-quality medicines. *Journal of Pharmaceutical Policy and Practice.* May 1, 13:9. doi: 10.1186/s40545-020-00211-9. PMID: 32377348; PMCID: PMC7193355.

25   World Health Professions Alliance. Be Aware: Tool for visual inspection of medicines. (2011). p. 4. Available: www.whpa.org/sites/default/files/2018-12/Too lkit_BeAware_Inspection.pdf

26   Mikomangwa, W. P., Madende, N. A., Kilonzi, M., Mlyuka, H. J., Ndayishimiye, P., Marealle, A. I., & Mutagonda, R. (2019). Unlawful dispensing practice of

diazepam: a simulated client approach in community pharmacies in the northwest of Dar-es-Salaam region, Tanzania. *BMC Health Services Research.* Aug 14, 19(1), 571. doi: 10.1186/s12913-019-4421-6. PMID: 31412944; PMCID: PMC6694629.

27   Chikowe, I., Bliese, S. L., Lucas, S., & Lieberman, M. (2018). Amoxicillin quality and selling practices in urban pharmacies and drug stores of Blantyre, Malawi. *American Journal of Tropical Medicine and Hygiene.* Jul, 99(1), 233–238. doi: 10.4269/ajtmh.18-0003. Epub 2018 Apr 19. PMID: 29692302; PMCID: PMC6085786.

28   Peyraud, N., Rafael, F., Parker, L. A., Quere, M., Alcoba, G., Korff, C., & Deats, M., et al. (2017). An epidemic of dystonic reactions in central Africa. *Lancet Global Health.* Feb, 5(2), e137–e138. doi: 10.1016/S2214-109X(16)30287-X. PMID: 28104176.

29   Kettler, H., White, K., & Hawkes, S. (2004). *Mapping the landscape of diagnostics for sexually transmitted infections: key findings and recommendations.* (UNICEF/UNDP/World Bank/WHO).

30   Land, K. J., Boeras, D. I., Chen, X S., Ramsay, A. R., & Peeling, R. W. (2019). REASSURED diagnostics to inform disease control strategies, strengthen health systems and improve patient outcomes. *Nature Microbiology.* Dec 13, 4, 46–54. doi: org/10.1038/s41564-018-0295-3

31   Kovacs, S., Hawes, S. E., Maley, S. N., Mosites, E., Wong, L., & Stergachis, A. (2014). Technologies for detecting falsified and substandard drugs in low and middle-income countries. *PLoS One.* Mar 26, 9(3):e90601. doi: 10.1371/journal.pone.0090601. PMID: 24671033; PMCID: PMC3966738.

32   Vickers, S., Bernier, M., Zambrzycki, S., Fernandez, F. M., Newton, P. N., & Caillet, C. (2018). Field detection devices forscreening the quality of medicines: a systematic review. *BMJ Global Health.* 3(4):e000725. doi: org/10.1136/bmjgh-2018-000725 PMID: 30233826

33   Roth, L., Biggs, K. B., & Bempong, D. K. (2019). Substandard and falsified medicine screening technologies. *AAPS Open.* 5(1):2.

34   Batson, J. S., Bempong, D. K., Lukulay, P. H., Ranieri, N., Satzger, R. D., & Verbois, L. (2016). Assessment of the effectiveness of the CD3+ tool to detect counterfeit and substandard anti-malarials. *Malaria Journal.* Feb 25,15:119. Doi: 10.1186/s12936-016-1180-2. PMID: 26917250; PMCID: PMC4766612.

35   Batson JS, Bempong DK, Lukulay PH, Ranieri N, Satzger RD, Verbois L. Assessment of the effectiveness of the CD3+ tool to detect counterfeit and substandard anti-malarials. *Malaria Journal.* 2016 Feb 25;15:119. doi: 10.1186/s12936-016-1180-2. PMID: 26917250; PMCID: PMC4766612.

36   Green, M. D., Mount, D. L., & Wirtz, R. A. (2001). Authentication of artemether, artesunate and dihydroartemisinin antimalarial tablets using a simple colorimetric method. *Tropical Medicine & International Health.* 6: 980–982. doi: org/10.1046/j.1365-3156.2001.00793.x

37   WHO, Basic Tests for Pharmaceutical Substances (1986), Basic Tests for Pharmaceutical Dosage Forms (1991), and Basic Tests for Drugs, Pharmaceutical Substances, Medicinal Plant Materials, and Dosage Forms (1998)

38   Kaale, E., Risha, P., & Layloff, T. (2011). TLC for pharmaceutical analysis in resource limited countries, *Journal of Chromatography A.* May, 1218(19), 2732–2736, doi: org/10.1016/j.chroma.2010.12.022.

39   Shewiyo, D. H., Kaale, E., Risha P. G., Dejaegher, B., Smeyers-Verbeke, J., & Van der Heyden, Y. (2012). HPTLC methods to assay active ingredients in pharmaceutical formulations: A review of the method development and validation steps, *Journal of Pharmaceutical and Biomedical Analysis*. 66:11–23, doi:org/10.1016/j.jpba.2012.03.034

40   World Health Organization. (1999). Counterfeit drugs: guidelines for the development of measures to combat counterfeit drugs. *World Health Organization*. Retrieved from https://apps.who.int/iris/bitstream/handle/10665/65892/WHO_EDM_QSM_99.1.pdf

41   Jähnke, R. W. O. & Dwornik, K. (2020). A concise quality control guide on essential drugs and other medicines: review and extension. 3rd edition. Global Pharma Health Fund.

42   Gnegel, G., Häfele-Abah, C., Neci, R.; Difäm-EPN Minilab Network, & Heide, L. (2022). Surveillance for substandard and falsified medicines by local faith-based organizations in 13 low- and middle-income countries using the GPHF Minilab. *Scientific Reports*. Jul 30, 12(1):13095. doi: 10.1038/s41598-022-17123-0. PMID: 35908047; PMCID: PMC9338985.

43   WHO. (2011). Survey of the quality of selected antimalarial medicines circulating in six countries of sub-Saharan Africa. Retrieved from www.afro.who.int/sites/defa ult/files/2017-06/WHO_QAMSA_report.pdf, accessed 3 April 2023.

44   Schäfermann, S., Hauk, C., Wemakor, E., Neci, R., Mutombo, G., Ngah Ndze, E., & Cletus, T., et al. (2020). Substandard and falsified antibiotics and medicines against noncommunicable diseases in western Cameroon and northeastern Democratic Republic of Congo. *American Journal of Tropical Medicine and Hygiene*. August, 103(2):894–908. doi: 10.4269/ajtmh.20-0184. Epub 2020 May 7. PMID: 32394884; PMCID: PMC7410427.

45   Risha, P. G., Msuya, Z., Ndomondo-Sigonda, M., & Layloff, T. P. (2006). Proficiency testing as a tool to assess the performance of visual TLC quantitation estimates. *Journal of AOAC International*, 89(5), 1300–1304 .

46   Yu, H., Le, H. M., Kaale, E., Long, K. D., Layloff, T., Lumetta, S. S., Cunningham, B. T. (2016). Characterization of drug authenticity using thin-layer chromatography imaging with a mobile phone. *Journal of Pharmaceutical and Biomedical Analysis*. Jun 5, 125:85–93. doi: 10.1016/j.jpba.2016.03.018. Epub 2016 Mar 9. PMID: 27015410.

47   Gnegel, G., Häfele-Abah, C., Neci, R., Difäm-EPN Minilab Network, & Heide, L. (2022). Surveillance for substandard and falsified medicines by local faith-based organizations in 13 low- and middle-income countries using the GPHF Minilab. *Scientific Reports*. Jul 30, 12(1):13095. doi: 10.1038/s41598-022-17123-0. PMID: 35908047; PMCID: PMC9338985.

48   Acevedo, A. J., Desai, D., Zaman, M. H., & Apiou-Sbirlea, G. (2022). PharmaChk: a decade of research and development towards the first quantitative, field-based medicine quality screening instrument. *Analyst*. 147, 3805–3816 doi: 10.1039/D2AN00284A

49   Yoshinori, M. (2016). Development of a simple and specific direct competitive ELISA for the determination of artesunate using an anti-artesunate polyclonal antiserum. *Tropical Medicine and Health*. Nov 21, 44(37). doi: 10.1186/s41182-016-0037-2.

50   Bartosh, A. V., Sotnikov, D. V., Hendrickson, O. D., Zherdev, A. V., & Dzantiev, B. B. (2020). Design of multiplex lateral flow tests: a case study for simultaneous detection of three antibiotics. *Biosensors* (Basel). Feb 27, 10(3):17. doi: 10.3390/bios10030017. PMID: 32120923; PMCID: PMC7146299.

51   CDC, Fentanyl Test Strips, a Harm Reduction Strategy. Retrieved from www.cdc. gov/stopoverdose/fentanyl/fentanyl-test-strips.html, accessed 19 Nov 2022

52   Guo, S., He, L., Tisch, D. J., Kazura, J., Mharakurwa, S., Mahanta, J., Herrera S., et al. (2016). Pilot testing of dipsticks as point-of-care assays for rapid diagnosis of poor-quality artemisinin drugs in endemic settings, *Tropical Medicine and Health*. 44, 15. doi: 10.1186/s41182-016-0015-8.

53   Lockwood, T. E., Vervoordt, A., Lieberman, M. (2021). High concentrations of illicit stimulants and cutting agents cause false positives on fentanyl test strips. *Harm Reduction Journal*. Mar 9, 18(1):30. doi: 10.1186/s12954-021-00478-4. PMID: 33750405; PMCID: PMC7941948.

54   Koesdjojo, M. T., Wu, Y., Boonloed, A., Dunfield, E. M., Remcho, V. T. (2014). Low-cost, high-speed identification of counterfeit antimalarial drugs on paper. *Talanta*, 130, (122–127). doi: 10.1016/j.talanta.2014.05.050

55   Noviana, E., Ozer, T., Carrell, C. S., Link, J. S., McMahon, C., Jang, I., & Henry C. S., (2021). Microfluidic Paper-Based Analytical Devices: From Design to Applications. *Chemical Reviews*. 121 (19), 11835–11885 doi: 10.1021/acs. chemrev.0c01335

56   Lisowski, P., Zarzycki, P. K. (2013). Microfluidic paper-based analytical devices (µPADs) and micro total analysis systems (µTAS): development, applications and future trends. *Chromatographia*. 76(19):1201–1214. doi: 10.1007/s10337-013-2413-y

57   Martinez, A. W., Phillips, S. T., Whitesides, G. M., & Carrilho, E. (2010). Diagnostics for the developing world: microfluidic paper-based analytical devices, *Analytical Chemistry*. 82 (1), 3–10, doi: 10.1021/ac9013989

58   Weaver, A. A., Reiser, H., Barstis, T., Benvenuti, M., Ghosh, D., Hunckler, M., Joy, B., et al. (2013). Paper analytical devices for fast field screening of beta lactam antibiotics and antituberculosis pharmaceuticals. *Analytical Chemistry*. Jul 2, 85(13):6453–60. doi: 10.1021/ac400989p. Epub 2013 Jun 18. PMID: 23725012; PMCID: PMC3800146.

59   PADs can be obtained from https://paperanalytics.org/ Accessed 20 Nov 2022.

60   Bannerjee et al; (2016). Visual recognition of paper analytical device images for detection of falsified pharmaceuticals, p. 437–445, *Proceedings of the IEEE Winter Conference on Applications of Computer Vision*. March 7–9, Lake Placid, NY.

61   Fu L-M & Wang Y-N. (2018). Detection methods and applications of microfluidic paper-based analytical devices. *TrAC Trends in Analytical Chemistry*. 107:196–211. doi: org/10.1016/j.trac.2018.08.018

62   Kim, T. H., Young, K. H., & Kim, M. S. (2020). Recent advances of fluid manipulation technologies in microfluidic paper-based analytical devices (µPADs) toward multi-step assays. *Micromachines*. 11(3), 269. doi: org/10.3390/mi11030269

63   Zambrzycki, S. C., Caillet, C., Vickers, S., Bouza, M., Donndelinger, D. V., Geben L. C., et al. (2021). Laboratory evaluation of twelve portable devicesfor medicine quality screening. *PLoS Negectedl Tropical Diseases* 15(9): e0009360.

64   Lockwood, T. E., Leong, T. X., Bliese, S. L., Helmke, A., Richard, A., Merga, G., Rorabeck, J., Lieberman, M., (2020). idPAD: paper analytical device for presumptive

identification of illicit drugs. *Journal of Forensic Science.* Jul, 65(4):1289–1297. Doi: 10.1111/1556-4029.14318. Epub 2020 Mar 30. PMID: 32227600; PMCID: PMC7332374.

65 Smith, M., Ashenef, A., & Lieberman, M. (2018). Paper analytic device to detect the presence of four chemotherapy drugs. *Journal of Global Oncology.* Dec, 4:1–10. doi: 10.1200/JGO.18.00198. PMID: 30589597; PMCID: PMC7010420.

66 Brooks, J. C., Mace, C. R. (2019). Scalable methods for device patterning as an outstanding challenge in translating paper-based microfluidics from the academic benchtop to the point-of-care. *Journal of Analyisis and Testing.* 3, 50–60. doi: org/10.1007/s41664-019-00093-0

67 Pollock, N. R., McGray, S., Colby, D. J., Noubary, F., Nguyen, H., Nguyen, T. A., Khormaee, S., et al. (2013). Field evaluation of a prototype paper-based point-of-care fingerstick transaminase test. *PLoS One.* Sep 30, 8(9):e75616. doi: 10.1371/journal.pone.0075616. PMID: 24098705; PMCID: PMC3787037.

68 Eberle, M. S., Ashenef, A., Gerba, H., Loehrer, P. J. Sr., Lieberman, M. (2020). Substandard cisplatin found while screening the quality of anticancer drugs from Addis Ababa, Ethiopia. *Journal of Global Oncology.* Mar, 6:407–413. doi: 10.1200/JGO.19.00365. PMID: 32142404; PMCID: PMC7113131.

69 Bliese, S. L., Maina, Mercy, Makoto Were, P., & Lieberman, M. (2019) Detection of degraded, adulterated, and falsified ceftriaxone using paper analytical devices. *Analytical Methods.* 11, 4727–4732

70 Gianini Morbioli, G., Mazzu-Nascimento, T., Stockton, A. M., & Carrilho, E. (2017). Technical aspects and challenges of colorimetric detection with microfluidic paper-based analytical devices (µPADs) – A review, *Analytica Chimica Acta.* 970:1–22 doi: org/10.1016/j.aca.2017.03.037

71 Soda Y. & Bakker E. (2019). Quantification of colorimetric data for paper-based analytical devices. *ACS Sensors* . 4 (12), 3093–3101 doi: 10.1021/acssensors.9b01802

72 Bannerjee et al,. (2016) Visual recognition of paper analytical device images for detection of falsified pharmaceuticals. pp437–445. *Proceedings of the IEEE Winter Conference on Applications of Computer Vision*, March 7–9, pp437–445. Lake Placid NY.

73 Acevedo, A. J., Desai, D., Zaman, M. H., & Apiou-Sbirlea, G. (2022). PharmaChk: a decade of research and development towards the first quantitative, field-based medicine quality screening instrument. *Analyst.* 147, 3805–3816. doi: 10.1039/D2AN00284A

74 Barstis, T. L. O, Flynn, P., & Lieberman, M. (2016). Analytical devices for detection of low-quality pharmaceuticals. US 9354181 B2, issued May 31,.

75 US Food and Drug Administration website, www.fda.gov/medical-devices/classify-your-medical-device/how-determine-if-your-product-medical-device, accessed 20 Nov 2022.

76 Batson, J. S., Bempong, D. K., Lukulay, P. H., Ranieri, N., Satzger, R. D., & Verbois, L. (2016). Assessment of the effectiveness of the CD3+ tool to detect counterfeit and substandard anti-malarials. *Malaria Journal.* 15:1.

77 Ranieri, N., Tabernero, P., Green, M. D., Verbois, L., Herrington, J., Sampson, E., Satzger, R. D., et al. (2014). Evaluation of a new handheld instrument for the detection of counterfeit artesunate by visual fluorescence comparison. *American*

Journal of Tropical Medicine and Hygiene. 91(5): 920–924. doi: 10.4269/ajtmh.13-0644

78   Jia, Y., Shelhamer, E., Donahue, J., Karayev, S., Long, J., Girshick, R., Guadarrama, S., & Darrell, T. (2014). Caffe: Convolutional architecture for fast feature embedding. arXiv preprint arXiv:1408.5093. doi: 10.48550/arXiv.1408.5093

79   Bowles P., Busch F. R., Leeman K. R., Palm A. S., & Sutherland K.. (2015). Confirmation of bosutinib structure; demonstration of controls to ensure product quality. *Organic Process Research & Development.* 19 (12), 1997–2005 doi: 10.1021/acs.oprd.5b00244

80   Caillet, C., Vickers, S., Zambrzycki, S., Fernandez, F. M., Vidhamaly, V., Boutsamay, K., et al. (2021) A comparative field evaluation of six medicine quality screening devices in Laos. *PLoS Neglected Tropical Diseases* . 15(9): e0009674.

81   Newton, P. N., Schellenberg, D., Ashley, E A., Ravinetto, R., Green, M. D., Ter Kuile, F. O., Tabernero, P., et al. (2015) Quality assurance of drugs used in clinical trials: proposal for adapting guidelines. *British Medical Journal.* Feb 25, 350:h602. doi: 10.1136/bmj.h602. PMID: 25716700; PMCID: PMC6705347.

82   Caillet, C., Vickers, S., Zambrzycki, S., Luangasanatip, N., Vidhamaly, V., Boutsamay, K., et al. (2021). Multiphase evaluation of portable medicines quality screening devices. *PLoS Neglected Tropical Diseases.* 15(9): e0009287

83   Vickers, S., Bernier, M., Zambrzycki, S., Fernandez, F. M., Newton, P. N., & Caillet, C. (2018). Field detection devices for screening the quality of medicines: a systematic review. *BMJ Global Health.* 3(4):e000725. pmid:30233826

84   Zambrzycki, SC, Caillet, C, Vickers, S, Bouza, M, Donndelinger, DV, Geben, LC, et al. (2021)Laboratory evaluation of twelve portable devicesfor medicine quality screening. *PLoS Neglected Tropical Diseases.* 15(9): e0009360.

85   Caillet, C., Vickers, S., Zambrzycki, S., Fernandez, F. M., Vidhamaly, V., Boutsamay., K, et al. (2021). A comparative field evaluation of six medicine quality screening devices in Laos. *PLoS Neglected Tropical Diseases.* 15(9): e0009674.

86   Luangasanatip, N., Khonputsa, P., Caillet, C., Vickers, S., Zambrzycki, S., Ferna´ndez, F. M., et al. (2021). Implementation of field detection devices for antimalarial quality screening in Lao PDR—A cost-effectiveness analysis. *PLoS Neglected Tropical Diseases.* 15(9): e0009539.

87   Caillet, C., Vickers, S., Vidhamaly, V., Boutsamay, K., Boupha, P., Zambrzycki, S., et al. (2021). Evaluation of portable devices for medicine quality screening: lessons learnt, recommendations for implementation, and future priorities. *PLoS Medicine* 18(9): e1003747

88   USP General Chapter <1850>Evaluation of Screening Technologies for Assessing Medicine QualityReprinted from USP42-NF37 (Official as of 1-Nov-2020)

89   U.S. Pharmacopeia (2020). USP Technology Review: Paper Analytical Device (PAD). TheTechnology Review Program. Rockville, Maryland.

90   Bannerjee et al., (2016) Visual Recognition of Paper Analytical Device Images for Detection of Falsified Pharmaceuticals, *Proceedings of the IEEE Winter Conference on Applications of Computer Vision*, March 7–9, Lake Placid, NY.

91   PADreader, available at https://play.google.com/store/apps/details?id=edu.nd.crc. paperanalyticaldevices, accessed 20 Nov 2022.

92 Paper Analytical Device Reader v 1.1, available at the Apple App Store, accessed 1 Dec 2022.

93 Hayes, K., Meyers, N., Sweet, C., Ashenef, A., Johann, T., Lieberman, M., & Kochalko, D. (2022). Securing the chain of custody and integrity of data in a global north-south partnership to monitor the quality of essential medicines. *Blockchain in Healthcare Today*, 5(S1). doi: org/10.30953/bhty.v5.230

2 HPLC for Analysis of Substandard and Falsified Medicinal Products

Eric Deconinck[1]*and Celine Vanhee[1]
[1]Scientific Direction Chemical and Physical Health Risks, Service of Medicines and Health Products, Sciensano, Brussels, Belgium

CONTENTS

2.1 Introduction 40
2.2 Falsified Medicines..41
    2.2.1 Illegal Preparations Containing Small Molecules 41
        2.2.1.1 Characterization of Suspected APIs...............................42
        2.2.1.2 Characterization of Suspected Finished Products............43
    2.2.2 Illegal Preparations Containing Macromolecules 48
        2.2.2.1 Protein or Peptide...50
        2.2.2.2 Use of LC-MS for the Identification of Illegal Peptide Drugs...50
        2.2.2.3 Use of LC-MS in the Identification of Illegal Protein Drugs...52
        2.2.2.4 Quantification of Polypeptide Drugs by Means of Chromatography 55
2.3 Regulated and Toxic Plants in Traditional Medicinal Products and Plant Food Supplements 55
2.4 Conclusion...61
References...63

2.1 INTRODUCTION

In the analysis of substandard and falsified (SF) medicinal products, liquid chromatography (LC) is used as the number one chromatographic technique, both in practice and literature. Liquid chromatography in its simplest form is thin layer chromatography (TLC). The basic principle is straightforward: solutions of the products under investigation are spot on a plate, coated with silica or some adapted version of it. Next, the plate is placed in a vessel conditioned with a solvent that runs over the plate, due to capillary forces. Depending on the affinity of the components under investigation toward the silica plate, migration will vary for different components allowing separation of mixtures and identification of reference solutions based on the positioning of the spots on the plate, and detected with either coloring agents, UV detection, or other modes. The major advantages of TLC is that it is easy to implement, cheap, and has already proved its value in the analysis of substandard and falsified medicines (Pachaly and Gesundheitshilfe Dritte Welt – German Pharma Health Fund 1994; Hadzija and Mattocks 1983; Hu et al. 2006; Moriyasu et al. 2001; Singh et al. 2009; Wu 2006; Pribluda et al. 2012) as well as offering a high performance version (Sheshashena Reddy et al. 2006; Shewiyo et al. 2009; 2011; Yemoa et al. 2017; Bhatt et al. 2016; Khuluza et al. 2016). However, TLC is often limited to targeted screening which is not compatible with screening for a wide spectrum of sometimes unknown medicinal products. High-pressure liquid chromatography (HPLC) with its separation on a chromatographic column packed with silica, or a modified form of it, and hyphenated with different detectors became the golden standard for the analysis of suspected products. The principle is similar to TLC, only here the stationary phase (most often a silica phase) is packed in a column, through which a mobile phase, consisting of a mixture of an aqueous phase (e.g., a buffer) and an organic phase is pumped, either at a constant composition (isocratic) or in a gradient. The samples are then injected into this system and the different components are separated based on their differences in affinity for the stationary and mobile phase. After separation, the components go to a detector. Although some other detectors can be found in literature – e.g., charged aerosol detection (Poplawska et al. 2014; 2013) and evaporative light scattering detection (Mutschlechner et al. 2018; Deconinck et al. 2013), the most commonly used detectors for the analysis of suspicious medicinal products are ultraviolet (UV) detection, single channel or diode array, and mass spectrometry (MS). HPLC hyphenated with UV/DAD is generally used for target analysis (presence of one or more known components) and as a quantification method, while the hyphenation with MS allows the untargeted screening of suspicious samples to reach an identification of the ingredient(s). Next to HPLC, ultra-high pressure chromatography (UHPLC) also offers a lot of advantages in the characterization of illegal and falsified medicines. UHPLC is based on the same principles as HPLC, but works with smaller silica particles in the stationary phase and using higher pressures. This allows faster analysis and thus higher throughput and resolution, and with less solvent consumption.

Liquid chromatography is applicable to a wide variety of suspected illegal medicinal products. The World Health Organization (WHO) uses the term "substandard and falsified (SF) medical products", which represents three mutually exclusive classes, namely substandard medicinal products, unregistered or unlicensed medicinal products, and falsified medicinal products (WHO 2018).

Substandard medicinal products are also called "out of specification" products, meaning these products are manufactured by regular companies, but the products show quality deficiencies and should be destroyed. Fraudulent practices, including theft, is the main raison why these products still enter the regular market. Unregistered or unlicensed medicines refer to products which are not approved for marketing by the national medicines regulatory authority (NMRA) of the market they are sold on. This means that these products may be legal in some countries but are illegal in others. Finally, falsified products deliberately or fraudulently misrepresent their identity, composition, or origin. Major issues with falsified medicines are having only inactive ingredients, wrong ingredients, or improper dosages (WHO 2018).

In addition to SF medicines, suspicious para-pharmaceutical products may also be encountered by customs, forensic, and control laboratories. An example of these "medicines in disguise" are products that are sold as dietary supplements or cosmetics, but contain one or more active pharmaceutical ingredient (API) (Deconinck et al. 2021; European Directorate for Quality of Medicines (EDQM) 2019; Pratiwi et al. 2021; Czepielewska et al. 2018) or regulated medicinal plant extracts (Deconinck, De Leersnijder, et al. 2013; Deconinck et al. 2019), potentially endangering consumers' health (Biesterbos et al. 2019). The discussion in this chapter will be structured according to the target molecules to be detected – small molecules, including low-molecular weight APIs used in allopathic medicine, and macromolecules, including polypeptide drugs and botanicals. The third section, before concluding the chapter, discusses detection methods for regulated and toxic plant extracts used in medicinal products and food supplements.

2.2　FALSIFIED MEDICINES

2.2.1　ILLEGAL PREPARATIONS CONTAINING SMALL MOLECULES

Falsifications of blockbuster drugs have been encountered on a worldwide scale for several decades, but there are regional differences in the products targeted by criminals and criminal organizations. Examples of these are the high occurrence of lifesaving medicines like antibiotics and antimalarial products in African countries. In Western countries, although they occur, falsified lifesaving medicines are less frequently encountered and the focus of falsifiers is on lifestyle medicines like enhancers for sexual and sport performance, and weight loss. These SF products often contain registered APIs – i.e., active ingredients already present in recognized medicinal products – or designer molecules. These are based on the registered molecule, recognized as medicine, but structural modifications are applied to avoid detection. A well-known example is the series of designer analogues of the PDE-5 inhibitors, mainly for sildenafil citrate and tadalafil (B.J. Venhuis and de Kaste 2012; Reeuwijk et al. 2013; Lee et al. 2021). Other designer molecules – such as the anorexic sibutramine, forbidden in Europe and the US – have also been encountered (Yun et al. 2018; Skalicka-Woźniak, Georgiev, and Orhan 2017). These designer molecules are non-INN (international nonproprietray names) molecules that were never recognized to be used as a medicine or were developed as medicines, but failed clinical trials. A Pan-European study was conducted on these molecules and showed that the top five categories of non-INN drugs encountered were anabolics, research peptides, PDE-5 inhibitors, selective androgen receptor modulators (SARMs), and drugs targeting the central nervous system (Deconinck et al. 2021).

Concerning the analytical characterization of these small molecules, two situations will be discussed. The first is the falsification of APIs sold to be processed and used in the manufacturing of medicinal products, and, second, the screening and qualitative and quantitative analysis of medicines and medicines in disguise.

2.2.1.1 Characterization of Suspected APIs

The availability of substandard and falsified raw materials is an often-neglected problem. Any form of tampering or falsification that affects the quality of APIs constitutes a direct threat to the health of patients. Several prominent cases increased the awareness that falsified APIs are a significant threat, resulting in a full program for the Generalised European Official Medicines Networks (GEON) API-working group to tackle this problem at a European level (Rebiere et al. 2022; Deconinck et al. 2022). The most well-known cases of API falsification are related to heparin and glycerin, where the pharmaceutical ingredients were intentionally contaminated with large amounts of relatively cheap and toxic substances (Beyer et al. 2010). The added substances escaped detection during routine analytical tests, which resulted in a large number of casualties (Holzgrabe and Malet-Martino 2011; Labadie 2012). Apart from falsification issues, inspections also regularly reveal serious non-compliance issues with good manufacturing practices (GMP) at certain API manufacturing sites, but also in the supply chain, that endanger the quality of the finished products and thus potentially also patients' health.

From an analytical point of view, the characterization of these suspected APIs can be analyzed according to monographs in respective pharmacopoeias, like the European Pharmacopoeia (Ph. Eur.) (Council of Europe 2020) or the US States Pharmacopoeia (USP) (US Pharmacopoeia Convention 2021). These monographs are elaborate and include multiple analytical methodologies in order to check the compliance and the quality of the products and not for the identification of falsifications. In other words, compliance in a pharmacopoeia does not automatically mean that the API is genuine, or that it is without risk, since the tests of the monograph may not reveal some aspects as they were not designed for it. An example of this is the "sartan crisis," where nitrosamines were detected in APIs and the finished products (EDQM 2021). Also, designer APIs mimicking raw materials are sometimes encountered by controlling agencies.

To the authors' knowledge, the only structured initiative to tackle this problem is the API fingerprint program of the GEON at the European level. In this program, studies are focused on an API molecule selected based on its susceptibility to be falsified (e.g. blockbuster drugs or expensive lifestyle medicines) or reported incidents in one of the member states of the network. The idea is to select a number of analytical methods/techniques to create a "fingerprint" of an API product, allowing distinction of the samples by origin (e.g., manufacturer, production site), verification of authenticity of suspect samples, and successful detection of falsified APIs. The analytical techniques for these studies are selected according to the nature of the selected molecule, using the tests of the respective monograph of the Ph. Eur. as a starting point. Three most-often selected pharmacopeial tests are infrared spectroscopy and liquid chromatography for related substances and residual solvents analyses with gas chromatography. These tests are then complemented with analytical techniques known to be of interest in "fingerprinting". The selection of analytical techniques should provide complementary data on structure (spectroscopy), organic impurities (related substances, residual solvents, NMR) and inorganic impurities (XRPD). To make the distinction between the different manufacturers, unsupervised chemometric techniques are applied. These studies were limited to the use of principal component analysis (PCA) and hierarchical clustering (Massart 1997). This is to ensure that the differentiation is solemnly due to differences in analytical data and so chemical differences between the samples (Rebiere et al. 2022; Deconinck et al. 2022).

Liquid chromatography as the method for detecting related substances is a very important test, not only for quality compliance, but also in the detection of falsified APIs. The mentioned studies were performed only on genuine products, which limited the ability of liquid chromatography to make distinctions among the origins of the samples, though it is to be expected that falsified APIs are of lower quality, containing more related substances or impurities due to non-GMP manufacturing and less thorough purification steps. More analyses of the impurities in falsified API samples could give an indication on the synthesis route, revealing not only patent infringements, but also allowing to relate different samples to one (illicit) origin, which is important in forensic cases. In that context, the analyses of impurities and related substances using liquid chromatography hyphenated with mass spectrometry (LC-MS) could reveal more information. The data generated by LC-MS could also increase the performance of liquid chromatography and related substance analysis in fingerprint studies, allowing distinctions based on origin, authenticity checks, and fast identification of suspect samples. The systematic screening of suspected API samples with LC-MS and/or GC-MS could also lead to the early detection of quality deficiencies, as was the case during the sartan crisis (EDQM 2021).

2.2.1.2 Characterization of Suspected Finished Products

In contrast to raw material products, finished products generally consist of a more complex matrix, composed of a mixture of excipients and one or several APIs or designer molecules. In the case of medicines in disguise (e.g. dietary supplements) the matrix can even be more complex (e.g. a mixture of dried herbs and plants or a cosmetic matrix). In these cases, the separation of the different components of the products is mandatory in order to fully characterize it and to identify the present APIs, and possibly toxic contaminants or impurities. Liquid chromatography is the golden standard for analytical separation for several purposes: targeted analysis (presence of one or a series of known or predefined compounds), quantification analysis, and, when hyphenated with mass spectrometry, untargeted screening. Although some other approaches – e.g., based on nuclear magnetic resonance (NMR) or spectroscopy – exist, the general approach of dealing with a suspected medicinal product entails untargeted screening using chromatography (gas and/or liquid) hyphenated with high resolution MS or MS/MS. In addition, quantitation of the found API or designer molecule can also be performed by LC-MS or LC-UV (Johansson et al. 2014; Vanhee et al. 2018). This is also reflected by the official document of the Illegal Medicines Working group of the GEON network. They created an "aide memoire" giving guidance for the analysis of suspected, falsified and illegal medicines (GEON 2019). As an example, Figure 2.1 shows the "screening protocol" of the aide memoire. (For other schemes offering guidance for more specific products we refer to GEON 2019). This clearly indicates that liquid chromatography plays an important role in the strategy, both in the screening (LC-MS) and for quantification.

The fact that LC-MS, next to targeted analysis, allows the untargeted screening of unknown samples for the presence of chemical drug compounds and, when an unknown compound is encountered, in combination with other techniques (e.g., NMR and IR), an unambiguous identification and structure elucidation of a previously unknown compound, makes it the method of choice to deal with suspicious products and preparations. For LC-MS, a clear distinction should be made between the different MS technologies routinely utilized by controlling laboratories – i.e., quadrupole, ion-trap, and Time-of-Flight (ToF). Liquid chromatography hyphenated with a quadrupole or a triple quadrupole MS is often only used for targeted analysis and is less suited for the screening of suspected illegal products and preparations. Ion trap MS allows screening of products based on the MS^1 spectrum, showing the precursor ion and often allowing to determine the estimated molecular mass of a molecule and the MS^2 fragmentation pattern that can be compared to the MS^2 spectra present in a library. Also, further fragmentation (MS^n) can be used, but they are generally less important in the detection of small molecules. Next to the ion trap, a high resolution accurate mass version, called orbitrap, exists. This technique allows the determination of the exact mass of a molecule (sub-ppm mass error), as well as the HRAM fragmentation patterns, allowing library searches. Identification of molecules is therefore possible based on both the exact mass and the fragmentation. Also the ToF detectors and the combination of a tandem quadrupole (Q-ToF) belong to the high resolution MS technologies (> 5 ppm mass error), allowing the determination of exact mass and identification of the molecules based on elemental analysis. With ToF and Q-ToF instruments a kind of fragmentation pattern, called MS^E, can be obtained but this

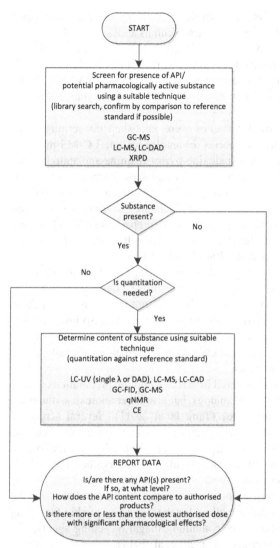

FIGURE 2.1 General screening and analysis protocol as described in the document of the European GEON network.

Reprinted from Generalised European OMCL Network (GEON 2019).

is generally considered less performant compared to the patterns obtained with either collision induced (CID) MS^2 using an ion trap and or higher-energy C-trap dissociation (HCD) applied with orbitrap technology. The parameters of these MS detectors (ion trap, orbitrap and (Q)ToF) can be chosen in such way that a large range of molecules and molecular masses can be detected.

In the context of PDE-5 inhibitors, a whole series of papers were published detecting new unregistered analogs or impurities (Venhuis and de Kaste 2012). The majority was detected using LC-MS and identified using a combination of MS, NMR and IR. Several LC-MS screening methods for these PDE-5 inhibitors, their analogues, and impurities have been described (Gratz, Flurer, and Wolnik 2004; Zou et al. 2006; Gryniewicz et al. 2009; Zhu et al. 2005; Fleshner et al. 2005; Lee et al. 2021; 2019).

Besides PDE-5 inhibitors, similar studies were published concerning other types of molecules or therapeutic categories. Examples include LC-MS methods to detect falsified and substandard sulphadoxine-pyrimethamine and amiodaquine products, first-line medicines in the treatment of malaria (Amin et al. 2005; Chutvirasakul et al. 2021), a LC-MS screening method for the qualitative analysis of alpha-glucosidase inhibitors in suspected products sold to treat diabetes (Dai et al. 2010), a method to screen illegal antibiotic preparations found on the European market (Tie et al. 2019), and a method to detect tramadol in mislabeled or falsified products (Abdel-Megied and Badr El-Din 2019).

The detection of medicines in disguise can be challenging. These are products sold as dietary supplements, cosmetics, or even simple consumer goods, but contain APIs or derivatives of APIs. The challenge is not only that the products are more difficult to detect and intercept by the authorities, but also involve more complex matrices, augmenting the necessity of separation techniques in order to clearly separate matrix components, legal chemicals like vitamins, illegally added APIs and impurities, and contaminants.

A group of medicines often encountered in herbal slimming preparations are the anti-obesity drugs like sibutramine and analogs, but also other anorexics, diuretics, antidepressants, and laxative molecules (Tang et al. 2011). Several screening techniques, making use of LC-MS, were described to screen dietary supplements and herbal preparations for slimming purposes (Kim et al. 2009; Stypułkowska et al. 2011; B. J. Venhuis et al. 2011; Wang et al. 2008; Carvalho et al. 2011). Other examples of LC-MS screening methods for adulterated products are the ones described for the characterization of artesunate tablets, purchased on the Asian market (Hall et al. 2006) and a method for the screening of skin bleaching creams and other products suspected of containing illegal bleaching agents, like hydroquinone, tretinoin, and corticoids (Desmedt et al. 2014).

These methods all focus on a specific category of products, such as slimming preparations, antimalerials, or skin bleaching products. In practice, products have to be screened for a series of molecules from different therapeutical categories, since often no information is available on its origin or use/indication. The analyst is blind and the LC-MS screening methods should be as generically applicable as possible. Most control laboratories charged with screening of suspicious products use these kinds of general LC-MS screening methods, often using simple mobile phases, like water and acetonitrile with the addition of formic acid in concentrations of 0.01% to 0.1%, a generally applicable C18 column and a linear gradient starting at 99% to 95% water phase, until 90% to 95% organic phase. Only a few of these general screening methods were published. An example is the method presented by Bogusz et al., (Bogusz et al. 2006) allowing the screening of herbal supplements and traditional medicines for analgesics, antibiotics, antidiabetic drugs, antiepileptic drugs, aphrodisiacs, hormones and anabolics, psychotropics, and weight reducing drugs. A similar method, based on LC-linear ion trap MS was described by Chen et al., (Chen et al. 2009) and allows the screening of herbal products for blood pressure and lipid lowering agents, antidiabetics, weight reducing agents, and aphrodisiacs.

Although LC-MS can be the first method of choice theoretically, this technology is very expensive, both to purchase and operate, and not all control laboratories have this technology at their disposal. Liquid chromatography coupled to UV detection or diode array detection, can be a valuable alternative. The more affordable LC-UV/DAD is standard equipment in most laboratories charged with quality control of medicines and analysis of suspected medicinal or para-pharmaceutical products. Although an untargeted approach with UV detection is difficult, screening methods including the most occurring APIs or designer molecules in a certain therapeutic class are feasible, provided there is prior knowledge of the sample to analyze. The same is true for adulterants occurring in products with a specific indication like slimming, muscle building, or potency enhancement. In all these cases, the method has to be developed for a set of molecules of interest using classical method development and validation strategies. Although screening is limited to the molecules the methods were developed for, certainty of the results can never be 100% guaranteed. Moreover, most laboratories have procedures on how to deal with signals not corresponding in retention time and UV spectrum to the targeted molecules. In that case, further analysis or even outsourcing can be necessary to identify the compound in order to state a clear risk evaluation of the product. In this way, LC-UV or DAD can be part of a screening strategy, when LC-MS is not available at the laboratory itself, but that limited access – e.g., through collaborations with other laboratories – is an option. An example is the approach presented by Liu et al., (S. Y. Liu et al. 2001) which can screen herbal dietary supplements for 266 different pharmaceuticals. The strategy makes use of an LC-DAD combined with GC-MS screening, which seems to be a very performant alternative for laboratories with access to GC-MS but not LC-MS.

Next to its use as an alternative screening method, LC-UV or DAD is also the method of choice for the quantification of APIs and other adulterants in suspected para-pharmaceutical and medicinal products. Its stability, easiness of use and relatively low operating costs makes it more suited for the quantification of the compounds identified during screening, compared to semi-quantitative analysis with mass spectrometry. Of course this is dependent on the molecules found. If a non-UV absorbent molecule needs to be quantified, LC-MS or other techniques should be used.

Different methods were published for the analysis of suspected and illegal products. For PDE-5 inhibitors, most papers used an LC-UV or DAD method for the quantification of the registered APIs, though some of them were used as a limited screening method. An example is the method developed by Sacré et al., (Sacré et al. 2011) for the analysis of registered PDE-5 inhibitors and seven of the most occurring analogs and impurities. Several of the papers conducting quantitative analysis on illegal PDE-5 inhibitor preparations found products with doses lower than the therapeutic dose, but some reported overdosed products (Gratz, Flurer, and Wolnik 2004; Park and Ahn 2012; Sacré et al. 2011; Tomić et al. 2010). The latter products represent a huge risk to patients and clearly show that the detection of illegally added APIs in suspected products is not enough. After identification of the APIs or designer molecules, quantitative analysis is necessary to check the correct dose (wrong dosage can be an indication of falsification) and for risk evaluation and prevention. De Orsi et al. (De Orsi et al. 2009) developed an LC-DAD screening method for the determination of PDE-5 inhibitors, testosterone, and local anesthetics in cosmetic creams sold as erectile dysfunction remedies or to enhance female genital stimulation. Another study developed a LC-DAD method for the screening for Indian aphrodisiac ayurvedic/healthcare products for adulteration with PDE-5 inhibitors (Savaliya et al. 2010). Both of these latter methods clearly show the advantage of LC compared to, for example, spectroscopic methods – i.e., the applicability in the analysis of a wide spectrum of matrices, such as tablets, capsules, creams, herbal matrices, etc.

LC-UV methods were described for other therapeutic classes, most notably for antimalarial products (Amin et al. 2005; Gaudiano et al. 2006; Debrus et al. 2011) and antibiotics (Gaudiano et al. 2008; Tie et al. 2019), two medicine classes constituting pillars of health care in most African and other developing countries. Falsification of these products – with incorrect APIs, incorrect doses or the presence of toxic adulterants and impurities – can have lethal effects due to therapy inefficiency or adverse reactions as well as heighten the risk for the emergence of resistant strains, and therefore cannot be neglected.

Several methods were described in the domain of adulterated dietary supplements for screening and quantification. In this context, dietary supplements for weight control are by far the category on which most were published. Kim et al., (Kim et al. 2009) and Mikami et al. (Mikami et al. 2005) proposed some interesting screening methods. The first method is able to screen dietary supplements for a series of anti-diabetes and anti-obesity drugs, while the second is a method to screen for benzodiazepines. The latter are often added to dietary supplements for weight control to mask the side effects of the illegally added anorexics. Further, several LC-UV methods were proposed for the quantification of various adulterants often encountered in these kinds of dietary supplements (Stypułkowska et al. 2011; Almeida, Ribeiro, and Polese 2000; Deconinck et al. 2012).

2.2.2 Illegal Preparations Containing Macromolecules

Prior to 2008, the encounter of illegal polypeptide drugs by European regulatory agencies was rather a scarce phenomenon and was mainly limited to certain sport performance enhancers, such as falsified human growth hormone (hGH), human chorionic gonadotropin (hGC), and Erythropoietin (EPO). However, over the past 15 years, polypeptide drugs, encompassing a diverse set of peptide drugs and protein drugs were encountered more and more often by national medicines controlling agencies or laboratories in charge of doping controls. These types of drugs are for parental administration and are generally seized either as a lyophilized powder present in a vial or as syringes filled with liquid (Venhuis et al. 2016; Deconinck et al. 2021).

In the case of peptide and protein drugs, with the exception of lifesaving medications, these products can easily be acquired from illicit internet pharmacies, which are sometimes disguised as research companies. These peptide drugs are mainly sold to improve sports performance or to strive to comply to cultural and societal ideals, such as physical appearance enhancers (e.g. weight loss enhancers or tanning peptides), synthetic mental performance enhancers or nootropics, antiaging peptides, and so forth. The categories of proteins that are encountered by regulatory agencies include lifesaving medicines such as insulins and monoclonal antibodies (mAbs). However, in the case of insulins, although primarily known as lifesaving antidiabetic drugs, they are also sold and marketed on illicit forums for their anabolic properties (Anderson et al. 2018; Heidet et al. 2019). A summary of the most popular peptide drugs available through various suspected illicit internet pharmacies can be found in Table 2.1, while Table 2.2 describes the most popular protein drugs encountered by regulatory agencies. These products are either available through the same or similar channels as the peptide drugs or, in the case of mAbs, have been stolen by malignant entities for falsification purposes. A textbook example of such activity was demonstrated in 2014 with the case of Herceptin® from Roche. Several vials of this medicine containing trastuzumab were stolen in Italy and resurfaced in the UK, Finland, and Germany. Visual inspection of the suspected adulterated vials demonstrated that the batch numbers and expiry dates on most vials did not match those on the outer package. Moreover, according to the press release by European Medicines Agency (EMA) some of these vials, normally containing lyophilized powder, contained a liquid solution, while some vials showed evidence of tampering with the rubber stoppers, crimping caps, or lids (EMA 2014; Streit 2017).

Although the discrepancy of the packaging and a visual inspection of the vials was sufficient to deem the product had been tampered with, it stands to reason that a more in-depth analysis of these types of products are pivotal in order to prove that the suspected product contains either illegal peptide or protein drugs or the lack of active pharmaceutical ingredient/s when analyzing tampered legal protein drugs, including insulin and mAbs, which is a requirement for regulation and/or prosecution. Due to the diverse nature of polypeptide drugs, competent analytical laboratories have to resort to a plethora of different techniques to uncover the contents. Various electrophoretic, immunological, or mass spectrometry (MS) based techniques – or combinations of these techniques – are required to merely identify the API/s. Therefore, a distinction is made between the analytical methodologies deployed to identify peptides and proteins, those for the search for "usual suspects," and for newly emerging illegal polypeptide drugs.

2.2.2.1 Protein or Peptide

Prior to any analysis, some controlling laboratories opt to perform a rudimentary screening to check if any peptide or protein is present as is described in the position paper put forward by the OMCL network (GEON 2020). This can be done by the use of a dye binding assay such as Bradford (Høj et al. 2021) or sodium dodecyl sulphate – polyacrylamide gel electrophoresis (Janvier et al. 2017). The advantage of this electrophoresis technique is that in addition to the detection of polypeptides by a dye, it also allows a swift separation based on molecular weight. This migration pattern could harbor important information that can be used to (partially) unravel the identity of the polypeptide and allow controllers to make educated decisions regarding the subsequent identification strategy. In the case where a band is found in the lower molecular weight region, a direct analysis by mass spectrometry, whether or not coupled to chromatographic separation techniques, could be performed (Vanhee et al. 2015). Alternatively, in the case of the occurrence of distinct migration profiles, that might be typical to more a traditional falsified biologicals such as hGH, EPO, and hCG, a targeted analysis might be chosen, including immunological assays. However, when no distinct migration patterns are visible for the yet unknown polypeptide, a bottom-up analysis is generally performed. It might also be possible that no basic amino acids are present in the sequence and a negative result is obtained with SDS-PAGE and Coomassie blue staining (e.g., Epitalon peptide). Then, analysis via mass spectroscopic technologies can be employed to rule out the presence of a polypeptide or confirm the presence of small molecule therapeutics or contaminants.

2.2.2.2 Use of LC-MS for the Identification of Illegal Peptide Drugs

Liquid chromatography (LC) coupled to mass spectrometry (MS) is an indispensable tool in the identification of falsified peptide drugs. The chromatography generally takes place under acidic conditions so that the peptides harbor a positive charge, required for downstream analysis by the MS. Both high performance liquid chromatography (HPLC) and ultra-high performance liquid chromatography systems (UHPLC) utilizing classical C8 or C18 columns are frequently listed in different case-reports (see Table 2.1 and the references therein). Additionally, the use of C18 columns with a positive surface charge under acidic conditions have been utilized to analyze several of these compounds (Lauber et al. 2013; Vanhee, et al. 2015). These type of columns allowed sharper peaks with MS-compatible mobile phases, compared to the more classical C18 column chemistries. Also, high strength silica columns, designed to retain both polar (basic) and less polar substances under acidic conditions have shown good applicability to various illegal peptides (Vanhee et al. 2020). Alternatively, the application of hydrophilic interaction liquid chromatography (HILIC) either coupled to MS or to a diode-array detector (DAD) for respective identification or quantification of different illegal peptides, including some very small acid peptides, has been proven useful (Janvier et al. 2017).

In addition to UHPLC and HPLC systems, nano liquid chromatography (nanoLC) has been used, mainly by doping laboratories, since this type of chromatography results in higher sensitivity, which is often required when analyzing extracts from biological matrices.

As important as the chromatographic separation is the type of mass spectrometer and the type of approach – e.g., full scan or targeted. Targeted systems (e.g., triple-quadrupole mass spectrometers) are used to screen for the presence of known peptides, while full scan methodologies screen for the presence of one or multiple analytes with an m/z within the corresponding interval (e.g., 100–1400 m/z).

It stands to reason that the use of a full scan MS set-up is vital for the analysis of unknown illegal preparations for which no assumptions can be made regarding content. For routine purposes, the use of high resolution mass spectrometry (HRMS) is desired but not pivotal, provided that an appropriate fragmentation technique and robust chromatographic separation are being used. Indeed, the use of the parameter retention time – the m/z of the precursor ion and the occurrence fragment ions (corresponding to the amino acid sequence specific b or y ions) – might be sufficient to identify a peptide, provided that a reference standard was subjected to the same analysis as these fragments and their intensity are dependent on the sequence, type of instrument, and size and charge state of the precursor molecule. In general, collision induced dissociation (CID) fragmentation results in the cleavage of polypeptides between the amide bond of two adjacent amino acids. Fragments from the N-terminal side are defined in literature as b-type fragments, while fragments at the C-terminal side are defined as y-type fragments. Due to the perceptible fragmentation, namely alongside the peptide bond while generally preserving the functional group, CID is probably the most widely used fragmentation technique. However, also higher-energy C-trap dissociation or HCD, present in some trap-based systems (e.g., orbitrap), yield similar fragmentation patterns to CID but include peptide fragments (e.g., y1, b1, y2, b2, ...) containing information on the N terminal of the respective precursor and immonium ions, which are often lost with CID fragmentation in trap-based systems. The term HCD is not to be confused with higher energy CID fragmentation, often used in ToF-based systems employing collisions in the range of kiloelectron volt (keV). The term higher in HCD refers to the augmented radiofrequency voltage to confine ions in the C-trap. "conventional" low-energy CID fragmentation typically involves collisions below 100 eV.

In contrast to more routine analysis, all case reports describing the analysis of a novel peptide or peptide group employ high resolution mass spectrometry (HRMS) and subsequent fragmentation techniques (resulting in b and/or y ions) to achieve accurate identification. HRMS is necessary for accurate mass determination and the elucidation of isotopic patterns in the case where a multiple protonated bigger peptide is encountered that displays higher charge state envelopes. Furthermore, the addition of essential MS^2-experiments, enables a reconstitution of the amino acid sequence, based on either manual or *in silico* appointment of b/y-ions, a phenomenon termed *the novo* sequencing. This *the novo* sequencing generally results in one or several possible peptide sequences, which can then be verified by means of the injection of a custom synthesized reference standard for the unambiguous identification of the correct sequence.

2.2.2.3   Use of LC-MS in the Identification of Illegal Protein Drugs

In the case of "traditional" falsified proteins (e.g., EPO, hCG, and hGH) a first rough identification of the identity may be obtained by the application of electrophoretic techniques through comparison of the migration profiles. Afterward, when there is a strong indication of the presence of known traditional proteins, further analysis through means of immunological assays (dot blot, western blot, or ELISA) may suffice to identify falsified proteins and might even be the desired way to proceed for highly glycosylated proteins.

Reversed phase liquid chromatography is the most popular separation technique used in combination with mass spectrometric detection for the identification of the polypeptide present in a suspected illegal protein preparation. HPLC, UHPLC, and, in some cases, nano-LC were reported as LC-systems (see Table 2.2 and the references therein). Although the majority of these separation techniques are similar to the ones described for the analysis of peptides, HILIC is more often employed as a 2D-separation system with RPLC-systems for the analysis of polar peptides such as glycopeptides or phosphopeptides. Other separation techniques can be envisaged for the analysis of therapeutic proteins such as size-exclusion chromatography, ion-exchange chromatography, affinity chromatography, and hydrophobic interaction chromatography, although the majority of the established methodologies are rarely compatible with in-tandem MS analysis. Nevertheless, much efforts have been put into the development of MS compatible multidimensional LC methodologies, as exemplified by the methodology put forward in 2014 for the analysis of various insulins (Chambers et al. 2014).

For the analysis of proteins by means of MS, two major approaches are described in literature, namely the top-down approach (intact analysis) and the bottom-up approach (analysis of peptide fragments, obtained after induced distinct proteolysis mainly by proteases). Whether one approach is preferred over the other depends not only on the size and complexity of the respective polypeptide, but also on the capabilities of the MS-system in terms of type of MS-platform, mass accuracy, resolving power, fragmentation techniques, and available data-processing algorithm. One of the key aspects of both approaches is the distinct fragmentation of ionized polypeptides in smaller fragments, which, in turn, can be compared to *in silico* models, databases, or used for de novo sequencing (either manual or through algorithms) to elucidate the amino acid sequence, known as the primary structure of a protein.

Currently, the bottom-up analysis is still the preferred technique for the analysis of falsified protein drugs as it is applicable to a plethora of proteins. An important step in this process prior to LC-MS analysis, is the enzymatic digestion of the

TABLE 2.1

Summary of the most popular peptide drugs encountered by controlling agencies and available through illicit internet pharmacies

Usage	Main mode of action	Name	Identification methodology described in:
sports performance enhancers	growth hormone secretagogues	GHRP-2	(Thomas et al. 2010; Kohler et al. 2010; Thomas, Höppner, et al. 2011; Maria Cristina Gaudiano, Valvo, and Borioni 2014; Vanhee, Janvier, et al. 2015; Hullstein et al. 2015; Høj et al. 2021);
		Gly-GHRP-2	(Krug et al. 2018; Popławska and Błażewicz 2019; Gajda et al. 2019)
		GHRP-6	(Thomas, Höppner, et al. 2011; Vanhee, Janvier, et al. 2015; Hullstein et al. 2015; Høj et al. 2021)
		Gly-GHRP-6	(Krug et al. 2018; Gajda et al. 2019)
		Ipamorelin	(Thomas, Höppner, et al. 2011; Semenistaya et al. 2015; Vanhee, Janvier, et al. 2015; Hullstein et al. 2015)
		Gly-Ipamorelin	(Krug et al. 2018; Gajda et al. 2019)
		Hexarelin	(Thomas, Höppner, et al. 2011; Semenistaya et al. 2015)
		GHRH	(Knoop et al. 2016)
		Tesamorelin	(Thevis, Thomas, and Schänzer 2011)
		GRF 1-29	(Vanhee, Janvier, et al. 2015; Høj et al. 2021; Thomas et al. 2012)
		Mod GRF (1-29)	(Gajda et al. 2019)
		Gly-Mod GRF (1-29)	(Gajda et al. 2019)
		CJC1295	(Henninge et al. 2010; Thomas et al. 2012; Vanhee, Janvier, et al. 2015; Hullstein et al. 2015)
		CJC1295-DAC	(Vanhee, Janvier, et al. 2015; Hullstein et al. 2015)

(*continued*)

TABLE 2.1 (Continued)
Summary of the most popular peptide drugs encountered by controlling agencies and available through illicit internet pharmacies

Usage	Main mode of action	Name	Identification methodology described in:
		Grehlin	(Eslami, Ghassempour, and Aboul-Enein 2017; Thomas et al. 2021)
		Gonadorelin	(Thomas et al. 2008; Zvereva, Dudko, and Dikunets 2018)
	insulin like growth factors	MGF	(Esposito, Deventer, and Eenoo 2012)
		PEG-MGF	(Temerdashev et al. 2017)
	corticosteroid release stimuli	ACTH 1-39	(Thomas et al. 2012)
		ACTH 1-24	(Thomas et al. 2012; Vanhee, Janvier, et al. 2015)
	EPO receptor agonist	ARA-290	(Judák et al. 2021)
	wound healing peptides	TB-500	(Ho et al. 2012; Esposito et al. 2012)
		Thymosine beta	(Vanhee, Janvier, et al. 2015; Hullstein et al. 2015)
		BPC-157	(Cox, Miller, and Eichner 2017; Høj et al. 2021)
Physical appearance enhancers	skin tanning peptides	Melanotan II	(Breindahl et al. 2015; Vanhee, Janvier, et al. 2015; Hullstein et al. 2015; Odoardi et al. 2021; Høj et al. 2021)
	weight loss peptides	Adipotide	(Vanhee, Janvier, et al. 2015)
		AOD9604	(Vanhee et al. 2014; Vanhee, Janvier, et al. 2015; Hullstein et al. 2015; Andersen Hartvig et al. 2014)
		hGH Fragment 176-191	(Vanhee et al. 2014)
	Anti-aging peptides	Epitalon	(Vanhee, Moens, et al. 2015; Vanhee, Janvier, et al. 2015)
Mental performance enchancers	nootropics	Selank	(Temerdashev et al. 2017; Vanhee et al. 2020)
		Semax	(Vanhee et al. 2020)
		Noopept	(Vanhee et al. 2020)
		Kisspeptin10	(Z. Liu et al. 2013)
		oxytocin	(Janvier et al. 2017; Høj et al. 2021)
Sexual performance enhancer		Bremelanotide	(Vanhee, Janvier, et al. 2015; Odoardi et al. 2021; Mestria et al. 2021)

intact polypeptide into smaller fragments. To this end, multiple proteases have been utilized for the distinct cleavage of the polypeptide of interest, with trypsin (cleavage at the carboxyl site of the basic amino acids lysine or arginine when no proline succeeds these amino acids), Lys-C (cleavage at the carboxyl site of lysine), and Lys-N (cleavage at the amino site of lysine) frequently used (Giansanti et al. 2016). This digestion process can take place in solution or additional tools can be used to decrease interfering substances and increase the signal to noise ratio, such as molecular weight cut-off filters, immunoaffinity purification, and in-gel trypsin digestion.

The purpose of the subsequent LC-MS/MS analysis is to separate the different peptides that were generated prior to the ionization by means of the electrospray ionization (ESI) source for further downstream MS detection and fragmentation. In addition to CID, other fragmentation techniques such as HCD, electron transfer dissociation (ETD), and electron capture dissociation (ECD) are often employed or used in combination with each other to increase sequence coverage. Multiple protein identification algorithms also allow for the combination of spectra obtained via CID and HCD or CID and ETD to maximize the chance to find suitable candidates in different protein databases (e.g., UniProt). Another possibility in addition to protein database search is the *de novo* sequencing of the respective polypeptide. This approach was used to demonstrate the presence of a remnant polyhistidine tag to ease the purification process of the desired polypeptide while attached to the illegal polypeptide drug (Kohler et al. 2010).

2.2.2.4 Quantification of Polypeptide Drugs by Means of Chromatography

In addition to the identification of the polypeptide, the quantification of the identified polypeptide is also important. Concurrent with the quantification of small molecule drugs, LC-coupled to either MS or UV-detection can be employed. Possible quantification wavelengths are within the range of 210–220 nm (absorbance of peptide bond) and 254 nm and 280 nm (absorbance of aromatic wavelengths). UV-quantification of falsified samples was reported for multiple peptides (Vanhee et al. 2015; Janvier et al. 2017) and several proteins, including somatropin and insulins.

2.3 REGULATED AND TOXIC PLANTS IN TRADITIONAL MEDICINAL PRODUCTS AND PLANT FOOD SUPPLEMENTS

The popularity of plant food supplements and traditional herbal medicines, especially in the Western world, increased over the past decade, following the "return to nature" trend, where marketing campaigns exhaustively promote natural and traditional products. These products also got a boost due to the evolution toward self-medication, the questioning of allopathic medicines, and the misperception that natural and herbal are synonyms for "safe." As a consequence a whole market developed offering a wide range of products, representing high profits. The latter makes these products vulnerable for fraud and adulteration, especially since

TABLE 2.2
Summary of the most popular protein drugs encountered by controlling agencies and available through illicit internet pharmacies

Usage	Main mode of action	Name	Identification methodology described in:
sports performance enhancers	Insuline like growth factors	IGF-1 Des-1-3 IGF-1 R^3-IGF-1 LongR^3-IGF-1 IGF-2	(Thevis et al. 2009; Thomas, Kohler, et al. 2011; Thomas et al. 2012; Niederkofler et al. 2013; van den Broek et al. 2015; Andersen Hartvig et al. 2014; Vanhee et al. 2016; Thomas et al. 2017; Mongongu et al. 2021)
	Human growth hormone	Human growth hormone	(Thevis et al. 2009; 2014; Andersen Hartvig et al. 2014; van den Broek et al. 2015; Hullstein et al. 2015; Janvier et al. 2017)
	Follistatins	Follistatin-315 Follistatin-344	(Reichel, Gmeiner, and Thevis 2019)
	EPO	rhEPO's DPO PEGylated forms	(Yu et al. 2010; Reichel 2010; Qureshi et al. 2012; Marchand et al. 2020; Voss et al. 2021; Gajda et al. 2019)
	Erythropoiesis-stimulating agent	ACE-301 ACE-001 (Sotarcept) ACE-536 (Luspatercept)	(Reichel et al. 2018; Walpurgis et al. 2018; Lange et al. 2019)
Physical appearance enhancers	Botulinum toxins	BoNT A	(Hobbs et al. 2019)
Life-saving medicines	insulins	human insulin porcine insuline insulin glargine insulin glulisine insulin lispro insulin aspart insulin detemir insulin degludec proinsulin	(Thomas et al. 2012; van den Broek et al. 2015; Chambers et al. 2014; Vanhee et al. 2016; Thomas et al. 2021)
	mAbs	Adalimumab, Bevacizumab, infliximab, rituximab, trastuzumab,...	(Legrand et al. 2022; Generalised European OMCL network (GEON 2020)

they can also be purchased via internet, an enabling platform for the trade of falsified, suspicious, or illegal products (Kofi-Tsekpo 2004; Tshibangu et al. 2004; Miraldi et al. 2001; Félix-Silva et al. 2014; Mosihuzzaman and Choudhary 2008; (WHO 2000).

Two issues, resulting in possible health threats, may occur with plant food supplements and traditional herbal medicine, especially when they are purchased from an irregular source, like websites disclosing their identity: (a) a first problem can be that the supplement does not contain the plant or herb mentioned on the ingredient list. For this there are two possibilities: (1) the first is falsification or fraud, mainly motivated by economical profit, or (2) the second possibility is confounding. The therapeutic properties are often related to one specific plant species, while other related species have less or no therapeutic activities or have other, possible toxic properties. This confounding can be either deliberately (fraud, the other species is cheaper) or accidently due to confusion with, for example, nomenclature used in traditional Chinese or Aryuvedic medicine. (b) The second problem is adulteration. Also here, two different possibilities occur: chemical adulteration and herbal adulteration. In the first case, the supplement is spiked with a chemical drug or active pharmaceutical ingredient. These products falls under medicines in disguise and can be analyzed following the protocols put in place by the laboratory for the analysis of small molecules in finished products. In the second case, the product contains active plants or herbs which are not disclosed on the packaging, are regulated or even toxic.

The screening of herbal products for the presence of regulated and/or toxic forbidden botanics is a real challenge. Generally the quality control of herbs and plants is based on micro- and macroscopic analysis, though in the products targeted here, several plants are grinded, mixed together and compressed into tablets or capsules, which renders these type of visual analysis obsolete. Therefore molecular biological tools, using genetic markers, termed DNA metabarcoding, have gained use to unravel the composition of complex and processed herbal products. DNA metabarcoding have shown great potential as an open screening approach. Although this technology might pinpoint the presence of a certain plant genus or species, it does not confirm or deny the presence of a certain chemical that might be present in the plant or present in only parts of the plant. Therefore, often combinatory strategies are required to either detect the presence of a contamination with a toxic or regulated plant species or the detection of the desired plant. However, laboratories charged with the analysis of suspicious plant food supplements are often more chemical-oriented laboratories with no expertise or access to genetic analysis. Therefore, these analyses should be outsourced, which could represent a high cost for the laboratories involved. Here several approaches could be envisaged depending on the desired outcome. Either the food supplements could be screened for the presence of highly toxic herbal components by a more targeted approach such as LC-MS/MS (Kaltner et al. 2020). Alternatively, another LC-MS/MS screening strategy can also be envisaged when screening for certain specific desired plant metabolites (Ichim and Booker 2021). Additionally, liquid chromatography can be a used as valuable alternative by applying a chromatographic fingerprinting approach. This approach is generally based on a chromatographic method, which allows to separate the components present in an extract of the sample. The obtained chromatogram is then used as a characteristic profile or fingerprint to identify a certain plant by comparing it to a reference or data base. The latter approach is mainly used and is also officially recognized for the identification of crude plant material (Tistaert et al. 2011). Indeed, with crude plant material falsifications, confounding may occur and chromatographic fingerprinting can help to detect them. Chromatographic fingerprints on crude plant material were described for the identification of plants – or more precisely, related species (search for genus specific biomarkers) – for stability testing and quality control and for the classification of different samples according to their origin or harvesting time. For these purposes, chromatographic fingerprints recorded with liquid chromatography and a single wavelength UV detection were mainly used. Chromatographic fingerprinting using mass spectrometry was also described, but with a different goal – i.e., the prediction of pharmacological activities and the identification of potential active molecules in the plant – as the possible basis for a new medicine. Application of this approach to crude plant material and the chemometric and data interpretation tools necessary have already been reviewed (Tistaert et al. 2011; Kharbach et al. 2020). Despite the fact that chromatographic fingerprinting is well-established in the domain of crude plant materials and has its value in the identification of the correct plant and species, only limited literature and approaches are available that characterize herbal mixtures.

The authors group developed a structured approach for the identification of a targeted plant and even a limited screening for a few plants in herbal mixtures (Deconinck et al. 2013; Deconinck et al. 2015). The strategy consists of two steps. (1) A chromatographic method with extraction process is developed, based on an extract of a reference of the targeted plant. The method is optimized for the resolution of the different signals in the chromatogram and the number of peaks using experimental design. Signals were detected using one UV-wavelength, i.e., 254 nm, the general wavelength for the detection of aromatic compounds. (2) Triturations of the reference plant using various herbal matrices (negative for the targeted plant) are prepared in different proportions and a correlation analysis between the obtained chromatograms is performed. Based on the correlation, a threshold is selected at which the targeted plant has a high probability of presence in the mixture. An unknown sample would then be analyzed together with the extract of the pure plants and possibly some triturations. If the correlation is higher than the selected threshold, the targeted plant is considered present with high probability. The approach was tested with some real samples and its performance was verified with LC-MS – i.e., the comparison of the mass data of characteristic peaks occurring both in the extract of the reference as in the sample under investigation. This is the approach for a single plant, but if a compromise can be found for the chromatographic method, allowing to obtain a characteristic fingerprint for several plants, the approach would allow to screen herbal mixtures for these plants using one chromatographic run. The confirmation with LC-MS is not always mandatory, but in case of doubt, this could reinforce the decision and the final result. After the development of the approach with some common non regulated plants (Deconinck et al. 2013; Deconinck et al. 2015), it was applied in the context of herbal adulteration of plant food supplements with potency enhancement as indication (Custers et al. 2017) with following regulated plants: *Epimedium spp.* leaves, *Pausinystalia yohimbe* bark, and *Tribulus terrestris* fruit. Six suspected samples from which one claimed the presence of *Tribulus terrestris* on its label, while another claimed the presence of *Epimedium spp.*, were screened using the fingerprint approach. Figure 2.2 shows the correlation chart obtained for the three targeted plants and the prepared triturations.

Unfortunately, based on the UV fingerprints, the label claims and the presence of the regulated plants in the unknown samples could not be verified. This, due to the presence of a large signal in the samples. The fingerprints were recorded again, this time using MS/MS. The peak interfering with the UV fingerprint was identified as sildenafil and the MS fingerprints could clearly confirm the presence of *Tribulus terrestris* and *Epimedium spp.* in the respective samples, that claimed it. The major peaks of the fingerprint of the extract of the reference plant were also present in the fingerprint of the samples and their MS^2 spectra corresponded. As examples, the MS fingerprints of *Epimedium spp.* and the sample that claimed its presence are shown in Figure 2.3. In the other four samples, none of the three plants could be identified (Custers et al. 2017).

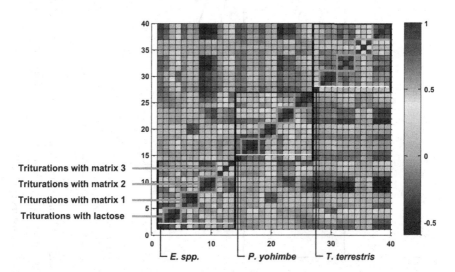

FIGURE 2.2 Correlation chart for the UV fingerprints for *Epimedium spp. Leaves, Pausinystalia yohimbe bark and Tribulus Terrestris fruit and their triturations.*

Reprinted from Custers et al. 2017.

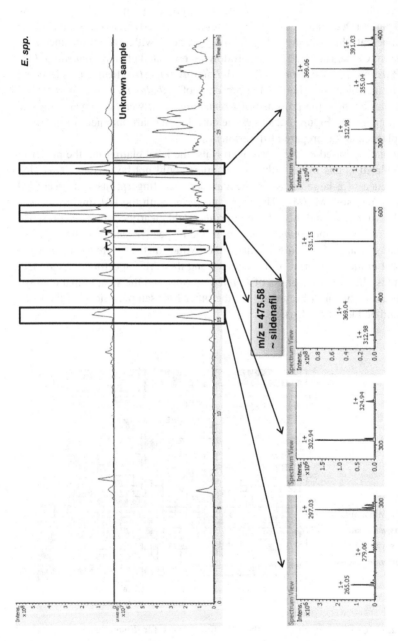

FIGURE 2.3 MS fingerpints of *Epimedium spp. Leaves* and an unknown sample claiming to contain *Epimedium spp.* The four most characteristic peaks of *Epimedium spp.* correspond both in retention times as mass data for reference and sample.

Adapted from Custers et al. 2017.

These results also confirm the power of LC-MS screening. Not only does it allow to identify the presence of targeted plants, but in the same run, also the possibly present chemical adulterants. All the experiments made use of LC-MS/MS based on ion trap technology, but the use of high resolution MS could increase the effectiveness of the approach and allow even more specific and characteristic fingerprints, allowing higher sensitivity and lower limits of detection for the targeted plants.

The failure of the UV fingerprints could be due to the interpretation of the fingerprints being based on visual inspection and correlation. It is known that the latter is highly influenced by interference of the other compounds and/or plants in the samples. Therefore in a follow-up study (Deconinck et al. 2019) the UV fingerprints were combined with chemometric modelling. The recorded fingerprints were used as a kind of "spectrum" to create binary models for each plant, meaning that for each plant a chemometric model was created allowing to distinguish between samples positive and negative for the targeted plant. Classical algorithms for this kind of binary models are k-nearest neighbors, soft independent modeling of class analogy (SIMCA) and partial least squares-discriminant analysis (PLS-DA). After validation of the models, the fingerprints of unknown samples can be used to classify the samples according to the presence or not of the targeted plant. This targeted approach was applied for the screening of 35 dietary supplements with slimming indication for *Aristolochia fanghi* and *Ilex paraguariensis* and 34 dietary supplements for potency enhancement for *Epimedium spp.*, *Pausinystalia yohimbe*, and *Tribulus terrestris*. The combined fingerprint-chemometric approach resulted in the identification of *Aristolochia fanghi* in one sample, *Ilex paraguariensis* in 11 samples, *Epimedium spp* in one sample, *Pausinystalia yohimbe* in six samples, and *Tribulus terrestris* in two samples (Deconinck et al. 2019). In order to validate the approach, the results were confirmed by repeating the analysis of the samples with LC-MS/MS.

This shows that the lower performance of relatively cheap LC-UV chromatography can be corrected using an additional analytical approach, in this case, chemometrics. Chemometrics is an analytical tool based on mathematics and is very cost effective to apply and to increase the results obtained with classical analytical tools like LC-UV or spectroscopic methods. The disadvantage is that investment is necessary in training in chemometrics in order to apply it correctly and allow a thorough validation of the created models. It is up to the laboratory, but if the investment in LC-MS equipment is not feasible, training in chemometrics could be a cheaper and valuable alternative with a broad range of possibilities.

2.4 CONCLUSION

From the discussion in this chapter it can be concluded that liquid chromatography hyphenated with MS, MS/MS, and UV plays a key role in the characterization of suspicious medicinal and para-pharmaceutical products. It has to be said that liquid chromatography, although pivotal, is not able to characterize and analyze the entire highly diverse range of products, control, customs, and forensic laboratories are dealing with. Indeed, for some products and purposes other analytical tools that will be explained in subsequent chapters could be more performant or efficient. An example of this are qPCR tests to characterize botanicals in herbal mixtures or gel electrophoresis with immunological detection for the identification of some biological molecules.

Next to the fact that other techniques are more advisable for certain types of products, it is often so that in laboratories dealing with these suspicious products a plethora of different techniques is used, taking advantage of their complementarity. For example in the analysis of biological products electrophoretic techniques are often combined with LC-MS, since they allow to first characterize the potentially present API and further to focus the procedures to follow in the LC-MS analysis. Another textbook example is the complementarity between LC-MS and GC-MS. True LC-MS is able to identify a broader range of pharmaceutically active molecules, though for some categories GC-MS is more specific and more suited. Examples here are volatile products like menthol and arnica, but also in the domain of the anabolics, and more specific the different analogs and esters of testosterone GC-MS allows a better, more specific and easier identification. The more sensitivities for certain molecules may differ between LC and GC-MS, making both techniques an ideal combination for a total screening of suspected samples. Also in the context of risk evaluation, the analysis of volatile impurities like residual solvents is important and often effectuated using GC-MS (Deconinck et al. 2013). Other complementarities can be found, like the combination of IR spectroscopy and mass spectrometry, for the identification of suspicious APIs or others between LC-MS and, for example, XRD or NMR.

Besides technical aspects and advantages, one should also think about resources and keep in mind that not all control laboratories worldwide have the same possibilities. For several laboratories in low and middle-income countries, the countries most affected by falsified and substandard medicines, the purchase, maintenance, and daily use of liquid chromatography, and more precisely LC-MS, is not self-evident. Indeed, LC-MS and certainly high-resolution MS is expensive to purchase, maintain, and use and necessitates controlled environmental conditions. This is probably the reason why alternatives are looked for using, for example, LC-UV or DAD methods as targeted or limited screening for products falling in certain therapeutic categories. Other alternatives exist, like the so called Quantum Dalton detector (QDA), which is a single quad mass spectrometric device with atmospheric pressure ionization. This is much cheaper and more robust (less influenced by environmental conditions). When hyphenated with a UHPLC and a DAD detector in series the detector allows a limited (targeted) screening of suspicious products based on retention, UV-spectrum and mass data. The mass data is mainly composed of the estimation of the molar mass and some limited fragmentation. Other detectors hyphenated with LC were explored for the analysis of suspicious pharmaceutical products, like evaporative light scattering detection

(ELSD), charged aerosol detection, or chemiluminescence, but the scope here is limited and therefore unsuited for the daily routine work in a control laboratory focused on suspicious medicinal products. They can be of use in the analysis of some products, but only after identification of the ingredients.

To conclude, liquid chromatography, especially hyphenated with MS and UV, is the golden standard in the characterization and analysis of suspicious medicinal products, but should fit in a complete strategy of a laboratory using a diversity of analytical tools in order to be able to characterize a broad range of products. This strategy should fit within the scope of the laboratory, the situation of the country, and the market they are dealing with, as well as whether necessary alternatives for LC-MS exist and can be used.

REFERENCES

Abdel-Megied, A. M., & Badr El-Din, K. M. (2019). Development of a novel LC-MS/MS method for detection and quantification of tramadol hydrochloride in presence of some mislabeled drugs: application to counterfeit study. *Biomedical Chromatography: BMC.* 33 (6): e4486. doi: org/10.1002/bmc.4486.

Almeida, A. E., Ribeiro, M. L., & Luciana Polese. (2000). Determination of amfepramone hydrochloride, fenproporex, and diazepam in so-called 'natural' capsules used in the treatment of obesity. *Journal of Liquid Chromatography & Related Technologies.* 23 (7): 1109–1118. doi: org/10.1081/JLC-100101512.

Amin, A. A., Snow, R. W., & Kokwaro, G. O. (2005). The quality of sulphadoxine-pyrimethamine and amodiaquine products in the Kenyan retail sector. *Journal of Clinical Pharmacy and Therapeutics.* 30 (6): 559–565. doi: org/10.1111/j.1365-2710.2005.00685.2.

Andersen Hartvig, R., Bjerre Holm, N., Weihe Dalsgaard, P., Reitzel, L. A., Breum Müller, I., & Linnet, K. (2012). Identification of peptide and protein doping related drug compounds confiscated in Denmark between 2007–2013. *Scandinavian Journal of Forensic Science.* 20 (2): 42–49. doi: org/10.2478/sjfs-2014-0003.

Anderson, L. J., Tamayose, J. M., & Garcia, J. M. (2018). Use of growth hormone, IGF-I, and insulin for anabolic purpose: pharmacological basis, methods of detection, and adverse effects. *Molecular and Cellular Endocrinology.* 464 (March): 65–72. doi: org/10.1016/j.mce.2017.06.010.

Beyer, T., Matz, M., Brinz, D., Rädler, O., Wolf, B., Norwig, J., Baumann, K., et al. (2010). Composition of OSCS-contaminated heparin occurring in 2008 in batches on the German market. *European Journal of Pharmaceutical Sciences.* 40 (4): 297–302. doi: org/10.1016/j.ejps.2010.02.002.

Bhatt, N. M., Chavada, V. D., Sanyal, M., & Shrivastav, P. S. (2016). Combining simplicity with cost-effectiveness: investigation of potential counterfeit of proton pump inhibitors through simulated formulations using thin-layer chromatography. *Journal of Chromatography. A.* 1473 (November): 133–142. doi: org/10.1016/j.chroma.2016.10.055.

Biesterbos, J. W. H., Sijm, D. T. H. M, Van Dam R., & Mol H. G. J. (2019). A health risk for consumers: the presence of adulterated food supplements in the Netherlands. *Food Additives & Contaminants. Part A, Chemistry, Analysis, Control, Exposure & Risk Assessment.* 36 (9): 1273–1288. doi: org/10.1080/19440049.2019.1633020.

Bogusz, M. J., Hassan H., Eid Al-Enazi, Z. I., & Al-Tufail, M. (2006). Application of LC-ESI-MS-MS for detection of synthetic adulterants in herbal remedies. *Journal of Pharmaceutical and Biomedical Analysis*. 41 (2): 554–562. doi: org/10.1016/j.jpba.2005.12.015.

Breindahl, T., Evans-Brown, M., Hindersson, P., McVeigh, J., Bellis, M., Stensballe, A., & Kimergård, A. (2015). Identification and characterization by LC-UV-MS/MS of melanotan II skin-tanning products sold illegally on the internet: melanotan II. *Drug Testing and Analysis*. 7 (2): 164–172. doi: org/10.1002/dta.1655.

Chambers, E. E., Fountain, K. J., Smith, N., Ashraf, L., Karalliedde J., Cowan D., & Legido-Quigley C. (2012). Multidimensional LC-MS/MS enables simultaneous quantification of intact human insulin and five recombinant analogs in human plasma. *Analytical Chemistry*. 86 (1): 694–702. doi: org/10.1021/ac403055d.

Chen, Y., Zhao, L., Lu, F. , Yu, Y., Chai, Y., & Wu, Y. (2009). Determination of synthetic drugs used to adulterate botanical dietary supplements using QTRAP LC-MS/MS. *Food Additives & Contaminants: Part A*. 26 (5): 595–603. doi: org/10.1080/02652030802641880.

Chutvirasakul, B., Joseph, J. F., Parr, M. K., & Suntornsuk, L. (2021). Development and applications of liquid chromatography-mass spectrometry for simultaneous analysis of anti-malarial drugs in pharmaceutical formulations. *Journal of Pharmaceutical and Biomedical Analysis*. 195 (February): 113855. doi: org/10.1016/j.jpba.2020.113855.

Council of Europe. (2020). *European Phamacopoeia 10th Edition*. 10th ed. Strasbourg, France: Council of Europe.

Cox, H. D., Miller, G. D., & Eichner, D. (2017). Detection and *in vitro* metabolism of the confiscated peptides BPC 157 and MGF R23H: detection and *in vitro* metabolism of BPC 157. *Drug Testing and Analysis*. 9 (10): 1490–98. doi: org/10.1002/dta.2152.

Custers, D., Van Praag, N., Courselle, P., Apers, S., & Deconinck, E.. (2017). Chromatographic fingerprinting as a strategy to identify regulated plants in illegal herbal supplements. *Talanta*. 164 (March): 490–502. doi: org/10.1016/j.talanta.2016.12.008.

Czepielewska, E., Makarewicz-Wujec, M., Różewski, F., Wojtasik, E., & Kozłowska-Wojciechowska, M. (2018). Drug adulteration of food supplements: a threat to public health in the European Union? *Regulatory Toxicology and Pharmacology*. 97 (August): 98–102. doi: org/10.1016/j.yrtph.2018.06.012.

Dai, X., Ning, A., Wu, J., Li, H., & Zhang, Q. (2010). Development and validation of HPLC-UV-MS method for the control of four anti-diabetic drugs in suspected counterfeit products. *Acta Pharmaceutica Sinica* 45 (3): 347–352.

De Carvalho, L. M., Martini, M., Moreira, A. P., De Lima, A. P. S., Correia, D., Falcão, T., Garcia, S. C., et al. (2010). Presence of synthetic pharmaceuticals as adulterants in slimming phytotherapeutic formulations and their analytical determination. *Forensic Science International*. 204 (1–3): 6–12. doi: org/10.1016/j.forsciint.2010.04.045.

De Orsi, D., Pellegrini, M., Marchei, E., Nebuloni, P., Gallinella, B., Scaravelli, G., Martufi, A., et al. 2009. High performance liquid chromatography-diode array and electrospray-mass spectrometry analysis of vardenafil, sildenafil, tadalafil, testosterone and local anesthetics in cosmetic creams sold on the internet web sites. *Journal of Pharmaceutical and Biomedical Analysis* 50 (3): 362–369. doi: org/10.1016/j.jpba.2009.05.022.

Debrus, B., Lebrun, P., Mbinze Kindenge, J., Lecomte, F., Ceccato, A., Caliaro, G., Mavar T. M. J., et al. (2011). Innovative high-performance liquid chromatography method development for the screening of 19 antimalarial drugs based on a generic approach, using design of experiments, independent component analysis and design space. *Journal of Chromatography. A.* 1218 (31): 5205–15. doi: org/10.1016/j.chroma.2011.05.102.

Deconinck, E., Canfyn, M., Sacré, P. Y., Courselle, P., & De Beer, J.O. (2013). Evaluation of the residual solvent content of counterfeit tablets and capsules. *Journal of Pharmaceutical and Biomedical Analysis.* 81–82 (July): 80–88. doi: org/10.1016/j.jpba.2013.03.023.

Deconinck, E., Courselle, P., Raimondo,M., Grange, Y., Rebière, H., Mihailova, A., Bøyum, O., et al. 2022. GEONs API fingerprint project: selection of analytical techniques for clustering of sildenafil citrate API samples. *Talanta.* 239 (March): 123123. doi: org/10.1016/j.talanta.2021.123123.

Deconinck, E., De Leersnijder, C., Custers, D., Courselle, P., & De Beer, J. O.. 2013. A strategy for the identification of plants in illegal pharmaceutical preparations and food supplements using chromatographic fingerprints. *Analytical and Bioanalytical Chemistry.* 405 (7): 2341–2352. doi: org/10.1007/s00216-012-6649-2.

Deconinck, E., Vanhamme, M., Bothy, J. L., & Courselle, P. (2019). A strategy based on fingerprinting and chemometrics for the detection of regulated plants in plant food supplements from the Belgian market: two case studies. *Journal of Pharmaceutical and Biomedical Analysis.* 166 (March): 189–196. doi: org/10.1016/j.jpba.2019.01.015.

Deconinck, E., Verlinde, K., Courselle, P., & De Beer, J. O.. (2012). A validated ultra high pressure liquid chromatographic method for the characterisation of confiscated illegal slimming products containing anorexics. *Journal of Pharmaceutical and Biomedical Analysis.* 59 (February): 38–43. doi: org/10.1016/j.jpba.2011.09.036.

Deconinck, E., Custers, D., and De Beer, J. O. (2015). Identification of (antioxidative) plants in herbal pharmaceutical preparations and dietary supplements. *Methods in Molecular Biology (Clifton, N.J.).* 1208: 181–199. doi: org/10.1007/978-1-4939-1441-8_12.

Deconinck, E., Vanhee, C., Keizers, P., Guinot, P., Mihailova, A., Syversen, P. V., Li-Ship, G., et al. (2021). The occurrence of non-anatomical therapeutic chemical-international nonproprietary name molecules in suspected illegal or illegally traded health products in Europe: a retrospective and prospective study. *Drug Testing and Analysis.* 13 (4): 833–840. doi: org/10.1002/dta.3001.

Desmedt, B., Van Hoeck, E., Rogiers, V., Courselle ,P., De Beer, J. O., De Paepe, K., & Deconinck, E. 2012. Characterization of suspected illegal skin whitening cosmetics. *Journal of Pharmaceutical and Biomedical Analysis.* 90 (March): 85–91. doi: org/10.1016/j.jpba.2013.11.022.

EDQM. (2021). The EDQM's response to nitrosamine contamination." European directorate for quality of medicines and health products. www.edqm.eu/en/edqms-response-nitrosamine-contamination.

Eslami, Z., Ghassempour, A., & Aboul-Enein, H. Y. (2017). Recent developments in liquid chromatography-mass spectrometry analyses of ghrelin and related peptides: recent developments in LC-MS analyses of ghrelin and related peptides. *Biomedical Chromatography.* 31 (1): e3796. doi: org/10.1002/bmc.3796.

Esposito, S., Deventer, K., & Van Eenoo, P. (2012). Characterization and identification of a c-terminal amidated mechano growth factor (MGF) analogue in black market products: c-terminal amidated MGF analogue in black market products. *Rapid Communications in Mass Spectrometry.* 26 (6): 686–692. doi: org/10.1002/rcm.6142.

Esposito, S., Deventer, K., Goeman, J., Van der Eycken, J., & Van Eenoo, P. (2012). Synthesis and characterization of the n-terminal acetylated 17-23 fragment of thymosin beta

4 identified in TB-500, a product suspected to possess doping potential: synthesis and characterization of AC-Tβ4(17-23). *Drug Testing and Analysis* 4 (9): 733–738. doi: org/10.1002/dta.1402.

European Directorate for Quality of Medicines (EDQM). (2019). Market surveillance of suspected illegal products (MSSIP) MSSIP004: medicines in disguise. EDQM. www. edqm.eu/sites/default/files/medias/fichiers/OMCL/study_report_on_medicines_in_d isguise.pdf.

European Medicines Agency (EMA). (2012). European Medicines Agency update on stolen vials of Herceptin. EMA. www.ema.europa.eu/en/documents/press-release/european-medicines-agency-update-stolen-vials-herceptin_en.pdf.

Félix-Silva, J., Brandt Giordani, R., Da Silva Jr., A. A., Zucolotto, S. M., & De Freitas Fernandes-Pedrosa, M. (2012). *Jatropha Gossypiifolia* L. (Euphorbiaceae): a review of traditional uses, phytochemistry, pharmacology, and toxicology of this medicinal plant. *Evidence-Based Complementary and Alternative Medicine.* 2014: 1–32. doi: org/10.1155/2014/369202.

Fleshner, N., Harvey, M., Adomat, H., Wood, C., Eberding, A., Hersey, K., & Guns, E. 2005. Evidence for contamination of herbal erectile dysfunction products with phosphodiesterase type 5 inhibitors. *The Journal of Urology.* 174 (2): 636–641; discussion 641; quiz 801. doi: org/10.1097/01.ju.0000165187.31941.cd.

Gajda, P. M., Bjerre Holm, N., Jakobsen Hoej, L., Schou Rasmussen, B., Weihe Dalsgaard, P., Ask Reitzel, L., & Linnet, K. (2019). Glycine-modified growth hormone secretagogues identified in seized doping material. *Drug Testing and Analysis.* 11 (2): 350–52. doi: org/10.1002/dta.2489.

Gaudiano, M. C., Di Maggio, A., Antoniella, E., Valvo, L., Bertocchi, P., Manna, L., Bartolomei, M., et al. 2008. An LC method for the simultaneous screening of some common counterfeit and sub-standard antibiotics validation and uncertainty estimation. *Journal of Pharmaceutical and Biomedical Analysis.* 48 (2): 303–309. doi: org/10.1016/j.jpba.2007.12.032.

Gaudiano, M. C., Antoniella, E., Bertocchi, P., & Valvo, L. (2006). Development and validation of a reversed-phase LC method for analysing potentially counterfeit antimalarial medicines. *Journal of Pharmaceutical and Biomedical Analysis.* 42 (1): 132–135. doi: org/10.1016/j.jpba.2006.01.059.

Gaudiano, M. C., Valvo, L., and Borioni, A. (2012). Identification and quantification of the doping agent GHRP-2 in seized unlabelled vials by NMR and MS: a case-report: identification of growth hormone releasing peptide-2 in not labelled vials. *Drug Testing and Analysis.* 6 (3): 295–300. doi: org/10.1002/dta.1603.

Generalised European OMCL network (GEON). (2019). PA/PH/OMCL (06) 81 R10: an 'aide-mémoire' for the testing of suspected illegally traded and falsified medicines. EDQM, Strasbourg, France. www.edqm.eu/sites/default/files/medias/fichiers/OMCL/Falsified_illegally_traded_medicines/omcl_aide_memoire_for_the_testing_of_suspected_illegally_traded_and_falsified_medicines_pa_ph_omcl_06_81_r10_january_2020.pdf.

———. (2020). PA/PH/OMCL (19) 93 R1: aide memoire: how to identify a falsified monoclonal antibody (MAb)/Ab containing fusion protein (AbFP)? EDQM, Strasbourg, France. www.edqm.eu/sites/default/files/medias/fichiers/OMCL/omcl_aide_memoire_how_to_identify_a_falsified_monoclonal_antibody_mab_ab_containing_fusion_protein_abfp_january_2020.pdf.

Giansanti, P., Tsiatsiani, L., Low, T. Y., & Heck, A. J. R. (2016). Six alternative proteases for mass spectrometry–based proteomics beyond trypsin. *Nature Protocols.* 11 (5): 993–1006. doi: org/10.1038/nprot.2016.057.

Gratz, S. R., Flurer, C. L., & Wolnik, K. A. (2002). Analysis of undeclared synthetic phosphodiesterase-5 inhibitors in dietary supplements and herbal matrices by LC-ESI-MS and LC-UV. *Journal of Pharmaceutical and Biomedical Analysis.* 36 (3): 525–533. doi: org/10.1016/j.jpba.2002.07.002.

Gryniewicz, C. M., Reepmeyer, J. C., Kauffman, J. F., & Buhse, L. F. (2009). Detection of undeclared erectile dysfunction drugs and analogues in dietary supplements by ion mobility spectrometry. *Journal of Pharmaceutical and Biomedical Analysis* 49 (3): 601–606. doi: org/10.1016/j.jpba.2008.12.002.

Hadzija, B. W., & Mattocks, A. M.. (1983). Simple Techniques to Detect and Identify Phentermine Adulteration. *Forensic Science International.* 23 (2–3): 143–147. doi: org/10.1016/0379-0738(83)90144-5.

Hall, K. A., Newton, P. N., Green, M. D., De Veij, M., Vandenabeele, P., Pizzanelli, D., Mayxay, M., et al. (2006). Characterization of counterfeit artesunate antimalarial tablets from Southeast Asia. *The American Journal of Tropical Medicine and Hygiene.* 75 (5): 804–811.

Heidet, M., Wahab, A. A., Ebadi, V., Cogne, Y., Chollet-Xemard, C., & Khellaf, M. (2019). Severe hypoglycemia due to cryptic insulin use in a bodybuilder. *The Journal of Emergency Medicine.* 56 (3): 279–81. doi: org/10.1016/j.jemermed.2018.10.030.

Henninge, J., Pepaj, M., Hullstein, I., and Hemmersbach, P. (2010). Identification of CJC-1295, a growth-hormone-releasing peptide, in an unknown pharmaceutical preparation. *Drug Testing and Analysis.* 2 (11–12): 647–50. doi: org/10.1002/dta.233.

Ho, E. N. M., Kwok, W. H., Lau, M. Y., Wong, A. S. Y., Wan, T. S. M., Lam, K. K. H., Schiff, P. J., and Stewart, B. D. (2012). Doping control analysis of TB-500, a synthetic version of an active region of thymosin β4, in equine urine and plasma by liquid chromatography–mass spectrometry. *Journal of Chromatography A.* 1265 (November): 57–69. doi: org/10.1016/j.chroma.2012.09.043.

Hobbs, R. J., Thomas, C. A., Halliwell, J., & Gwenin, C. D. (2019). Rapid detection of botulinum neurotoxins—a review. *Toxins.* 11 (7): 418. doi: org/10.3390/toxins11070418.

Høj, L. J., Schou Rasmussen, B., Weihe Dalsgaard, P., & Linnet, K. (2021). Analysis of seized peptide and protein-based doping agents using four complimentary methods: liquid chromatography coupled with time of flight mass spectrometry, liquid chromatography–ultraviolet, Bradford, and immunoassays. *Drug Testing and Analysis.* 13 (7): 1457–1463. doi: org/10.1002/dta.3026.

Holzgrabe, U., & Malet-Martino, M. 2011. Analytical challenges in drug counterfeiting and falsification—the NMR approach. *Journal of Pharmaceutical and Biomedical Analysis.* 55 (4): 679–687. doi: org/10.1016/j.jpba.2010.12.017.

Hu, C., Zou, W., Hu, W., Ma, X., Yang, M., Zhou, S., Sheng, J., et al. (2006). Establishment of a fast chemical identification system for screening of counterfeit drugs of macrolide antibiotics. *Journal of Pharmaceutical and Biomedical Analysis.* 40 (1): 68–72. doi: org/10.1016/j.jpba.2005.06.026.

Hullstein, I. R., Malerod-Fjeld, H., Dehnes, Y., & Hemmersbach, P. (2015). Black market products confiscated in Norway 2011–2014 compared to analytical findings in urine samples: black market products in Norway. *Drug Testing and Analysis.* 7 (11–12): 1025–1029. doi: org/10.1002/dta.1900.

Ichim, M. C. & Booker, A. (2021). Chemical authentication of botanical ingredients: a review of commercial herbal products. *Frontiers in Pharmacology.* 12 (April): 666850. doi: org/10.3389/fphar.2021.666850.

Janvier, S., De Sutter, E., Wynendaele, E., De Spiegeleer, B., Vanhee, C., & Deconinck, E. (2017). Analysis of illegal peptide drugs via HILIC-DAD-MS. *Talanta.* 174 (November): 562–571. doi: org/10.1016/j.talanta.2017.06.032.

Johansson, M., Fransson, D., Rundlöf, T., Huynh, N., and Arvidsson, T. (2012). A general analytical platform and strategy in search for illegal drugs. *Journal of Pharmaceutical and Biomedical Analysis.* 100 (November): 215–229. doi: org/10.1016/j.jpba.2012.07.026.

Judák, P., Esposito, S., Coppieters, G., Van Eenoo, P., & Deventer, K. (2021). Doping control analysis of small peptides: a decade of progress. *Journal of Chromatography B.* 1173 (May): 122551. doi: org/10.1016/j.jchromb.2021.122551.

Kaltner, F., Kukula, V., & Gottschalk, C. (2020). Screening of food supplements for toxic pyrrolizidine alkaloids. *Journal of Consumer Protection and Food Safety,* 15 (3): 237–243. doi: org/10.1007/s00003-020-01296-9.

Kharbach, M., Marmouzi, I., El Jemli, M., Bouklouze, A., & Vander Heyden, Y. (2020). Recent advances in untargeted and targeted approaches applied in herbal-extracts and essential-oils fingerprinting – a review. *Journal of Pharmaceutical and Biomedical Analysis.* 177 (January): 112849. doi: org/10.1016/j.jpba.2019.112849.

Khuluza, F., Kigera, S., Jähnke, R. W. O., & Heide, L. (2016). Use of thin-layer chromatography to detect counterfeit sulfadoxine/pyrimethamine tablets with the wrong active ingredient in Malawi. *Malaria Journal.* 15 (April): 215. doi: org/10.1186/s12936-016-1259-9.

Kim, S. H., Lee, J., Yoon, T., Choi, J., Choi, D., Kim, D., & Kwon, S. W. (2009). Simultaneous determination of anti-diabetes/anti-obesity drugs by LC/PDA, and targeted analysis of sibutramine analog in dietary supplements by LC/MS/MS. *Biomedical Chromatography: BMC.* 23 (12): 1259–1265. doi: org/10.1002/bmc.1248.

Knoop, A., Thomas, A., Fichant, E., Delahaut, P., Schänzer, W., & Thevis, M. (2016). Qualitative identification of growth hormone-releasing hormones in human plasma by means of immunoaffinity purification and LC-HRMS/MS. *Analytical and Bioanalytical Chemistry.* 408 (12): 3145–53. doi: org/10.1007/s00216-016-9377-3.

Kofi-Tsekpo, M. (2002). Institutionalization of African traditional medicine in health care systems in Africa. *African Journal of Health Sciences.* 11 (1–2): i–ii. doi: org/10.4314/ajhs.v11i1.30772.

Kohler, M., Thomas, A., Geyer, H., Petrou, M., Schänzer, W., & Thevis, M. (2010). Confiscated black market products and nutritional supplements with non-approved ingredients analyzed in the cologne doping control laboratory 2009. *Drug Testing and Analysis.* 2 (11–12): 533–537. doi: org/10.1002/dta.186.

Krug, O., Thomas, A., Malerød-Fjeld, H., Dehnes, Y., Laussmann, T., Feldmann, I., Sickmann, A., & Thevis, M. (2018). Analysis of new growth promoting black market products. *Growth Hormone & IGF Research.* 41 (August): 1–6. doi: org/10.1016/j.ghir.2018.05.001.

Labadie, J. (2012). Forensic pharmacovigilance and substandard or counterfeit drugs. *The International Journal of Risk & Safety in Medicine.* 24 (1): 37–39. doi: org/10.3233/JRS-2012-0551.

Lange, T., Walpurgis, K., Thomas, A., Geyer, H., & Thevis, M. (2019). Development of two complementary LC–HRMS methods for analyzing Sotatercept in dried blood spots for doping controls. *Bioanalysis.* 11 (10): 923–940. doi: org/10.4155/bio-2018-0313.

Lauber, M. A., Koza, S. M., McCall, S. A., Alden, B. A., Iraneta, P. C., & Fountain, K. J. 2013. High-resolution peptide mapping separations with MS-friendly mobile phases and charge-surface-modified C18. *Analytical Chemistry* 85 (14): 6936–6942. doi: org/10.1021/ac401481z.

Lee, J. H., Han, J. H., Kim, S., Kim, N. S., Yoon, C. , Kim, J., & Baek, S. Y. (2021). Screening sexual performance enhancing compounds and their analogues in counterfeit and illicit erectile dysfunction drugs by high-performance liquid chromatography and liquid chromatography-tandem mass spectrometry. *Journal of Forensic and Legal Medicine.* 82 (August): 102222. doi: org/10.1016/j.jflm.2021.102222.

Lee, J. H., Park, H. N., Park, O. R., Kim, N. S., Park, S., & Kang, H. (2019). Screening of illegal sexual enhancement supplements and counterfeit drugs sold in the online and offline markets between 2014 and 2017. *Forensic Science International.* 298 (May): 10–19. doi: org/10.1016/j.forsciint.2019.02.012.

Legrand, P., Dembele, O., Alamil, H., Lamoureux, C., Mignet, N., Houzé, P., & Gahoual, R. (2022). Structural identification and absolute quantification of monoclonal antibodies in suspected counterfeits using capillary electrophoresis and liquid chromatography-tandem mass spectrometry. *Analytical and Bioanalytical Chemistry*, January. doi: org/10.1007/s00216-022-03913-y.

Liu, S. Y., Woo, S. O. & Koh, H. L.. (2001). HPLC and GC-MS screening of Chinese proprietary medicine for undeclared therapeutic substances. *Journal of Pharmaceutical and Biomedical Analysis.* 24 (5–6): 983–92. doi: org/10.1016/s0731-7085(00)00571-9.

Liu, Z., Ren, C., Jones, W., Chen, P., Seminara, S. B., Chan, Y., Smith, N. F., et al. (2013). LC–MS/MS quantification of a neuropeptide fragment kisspeptin-10 (NSC 741805) and characterization of its decomposition product and pharmacokinetics in rats. *Journal of Chromatography B.* 926 (May): 1–8. doi: org/10.1016/j.jchromb.2013.02.027.

Marchand, A., Martin, L., Martin, J., Ericsson, M., & Audran, M. (2020). The case of the EPO-poisoned syringe. *Drug Testing and Analysis.* 12 (5): 637–640. doi: org/10.1002/dta.2757.

Massart, Desiré L., ed. 1997. *Handbook of Chemometrics and Qualimetrics.* Data Handling in Science and Technology, v. 20. Amsterdam; New York: Elsevier.

Mestria, S., Odoardi, S., Frison, G., & Strano Rossi, S. (2021). LC-HRMS characterization of the skin pigmentation and sexual enhancers melanotan ii and bremelanotide sold on the black market of performance and image enhancing drugs. *Drug Testing and Analysis* 13 (4): 876–882. doi: org/10.1002/dta.2986.

Mikami, E., Goto, T., Ohno, T., Oka, H., & Kanamori, H. (2005). Simultaneous analysis of seven benzodiazepines in dietary supplements as adulterants using high performance liquid chromatography and its application to an identification system for diazepam. *Journal of Health Science.* 51 (3): 278–283. doi: org/10.1248/jhs.51.278.

Miraldi, E., S. Ferri, and V. Mostaghimi. 2001. "Botanical Drugs and Preparations in the Traditional Medicine of West Azerbaijan (Iran)." *Journal of Ethnopharmacology* 75 (2–3): 77–87. https://doi.org/10.1016/s0378-8741(00)00381-0.

Mongongu, C., Coudoré F., Domergue, V., Ericsson, M., Buisson, C., and Marchand, A. (2021). Detection of long R 3 -IGF-I, Des(1-3)-IGF-I, and R 3 -IGF-I using immunopurification and high resolution mass spectrometry for antidoping purposes. *Drug Testing and Analysis.* 13 (7): 1256–1269. doi: org/10.1002/dta.3016.

Moriyasu, T., Shigeoka, S., Kishimoto, K., Ishikawa, F., Nakajima, J., Kamimura, H., & Yasuda, I. (2001). Identification system for sildenafil in health foods. *Yakugaku Zasshi: Journal of the Pharmaceutical Society of Japan.* 121 (10): 765–769. doi: org/ 10.1248/yakushi.121.765.

Mosihuzzaman, M., & Iqbal Choudhary, M. (2008). Protocols on safety, efficacy, standardization, and documentation of herbal medicine (IUPAC technical report). *Pure and Applied Chemistry.* 80 (10): 2195–2230. doi: org/10.1351/pac200880102195.

Mutschlechner, B., Schwaiger, S., Tran, T. V. A., & Stuppner, H. (2018). Development of a selective HPLC-DAD/ELSD method for the qualitative and quantitative assessment of commercially available Eurycoma longifolia products and plant extracts. *Fitoterapia.* 124 (January): 188–192. doi: org/10.1016/j.fitote.2017.11.015.

Niederkofler, E. E., Phillips, D. A., Krastins, B., Kulasingam, V., Kiernan, U. A., Tubbs, K. A., Peterman, S. M., et al. (2013). Targeted selected reaction monitoring mass spectrometric immunoassay for insulin-like growth factor 1. Edited by John Matthew Koomen. *PLoS ONE* 8 (11): e81125. doi: org/10.1371/journal.pone.0081125.

Odoardi, S., Mestria S., Biosa G., Valentini V., Federici S., & Strano Rossi S. 2021. An Overview on Performance and Image Enhancing Drugs (PIEDs) Confiscated in Italy in the Period 2017–2019. *Clinical Toxicology.* 59 (1): 47–52. doi: org/10.1080/ 15563650.2020.1770277.

Pachaly, Peter, and Gesundheitshilfe Dritte Welt – German Pharma Health Fund, eds. 1992. *Simple Thin Layer Chromatographic Identification of Active Ingredients in Essential Drugs.* Gesundheitshilfe Dritte Welt. Aulendorf: ECV – Editio-Cantor-Verl.

Park, M., & Ahn, S. (2012). Quantitative analysis of sildenafil and tadalafil in various fake drugs recently distributed in Korea: quantitative analysis of sildenafil and tadalafil. *Journal of Forensic Sciences.* 57 (6): 1637–1640. doi: org/10.1111/ j.1556-4029.2012.02162.2.

Popławska, M., & Błażewicz, A. (2019). Identification of a novel growth hormone releasing peptide (a glycine analogue of GHRP-2) in a seized injection vial. *Drug Testing and Analysis.* 11 (1): 162–167. doi: org/10.1002/dta.2467.

Poplawska, M., Blazewicz, A., Bukowinska, K., and Fijalek, Z. 2013. Application of high-performance liquid chromatography with charged aerosol detection for universal quantitation of undeclared phosphodiesterase-5 inhibitors in herbal dietary supplements. *Journal of Pharmaceutical and Biomedical Analysis.* 84 (October): 232–243. doi: org/10.1016/j.jpba.2013.06.018.

Poplawska, M., Blazewicz, A., Zolek, P., & Fijalek, Z. (2012). Determination of flibanserin and tadalafil in supplements for women sexual desire enhancement using high-performance liquid chromatography with tandem mass spectrometer, diode array detector and charged aerosol detector. *Journal of Pharmaceutical and Biomedical Analysis.* 94 (June): 45–53. doi: org/10.1016/j.jpba.2012.01.021.

Pratiwi, R., Fayza Dipadharma, R. H., Prayugo, I. J., & Layandro, O. A.. (2021). Recent analytical method for detection of chemical adulterants in herbal medicine. *Molecules (Basel, Switzerland.)* 26 (21): 6606. doi: org/10.3390/molecules26216606.

Pribluda, V. S., Barojas, A., Añez, A., López C. G., Figueroa, R., Herrera, R., Nakao, G., et al. (2012). Implementation of basic quality control tests for malaria medicines in Amazon Basin countries: results for the 2005–2010 period. *Malaria Journal.* 11 (June): 202. doi: org/10.1186/1475-2875-11-202.

Qureshi, Z. P., Norris, L., Sartor, O., McKoy, J. M., Armstrong, J., Raisch, D. W., Garg, V., Stafkey-Mailey, D., & Bennett, C. L. (2012). Caveat oncologist: clinical findings and consequences of distributing counterfeit erythropoietin in the United States. *Journal of Oncology Practice*. 8 (2): 84–90. doi: org/10.1200/JOP.2011.000325.

Rebiere, H., Grange, Y., Deconinck, E., Courselle, P., Acevska, J., Brezovska, K., Maurin, J., et al. (2022). European fingerprint study on omeprazole drug substances using a multi analytical approach and chemometrics as a tool for the discrimination of manufacturing sources. *Journal of Pharmaceutical and Biomedical Analysis*. 208 (January): 114442. doi: org/10.1016/j.jpba.2021.114442.

Reeuwijk, N. M., Venhuis, B. J., De Kaste, D., Hoogenboom, L. A. P., Rietjens, I. M. C. M., & Martena, M. J.. 2013. "Sildenafil and analogous phosphodiesterase type 5 (PDE-5) inhibitors in herbal food supplements sampled on the Dutch market. *Food Additives & Contaminants. Part A, Chemistry, Analysis, Control, Exposure & Risk Assessment*. 30 (12): 2027–2032. doi: org/10.1080/19440049.2013.848292.

Reichel, C. (2010). Practicing IEF-PAGE of EPO: the impact of detergents and sample application methods on analytical performance in doping control. *Drug Testing and Analysis*. 2 (11–12): 603–619. doi: org/10.1002/dta.215.

Reichel, C., Farmer, L., Gmeiner, G., Walpurgis, K., & Thevis, M. (2018). Detection of Sotatercept (ACE-011) in human serum by SAR-PAGE and western single blotting. *Drug Testing and Analysis*. 10 (6): 927–937. doi: org/10.1002/dta.2346.

Reichel, C., Gmeiner, G., & Thevis, M. (2019). Detection of black market Follistatin 342. *Drug Testing and Analysis*. 11 (11–12): 1675–1697. doi: org/10.1002/dta.2741.

Sacré, P., Deconinck E., Chiap P., Crommen J., Mansion F., Rozet E., Courselle P., & De Beer J. O. 2011. Development and validation of a ultra-high-performance liquid chromatography-UV method for the detection and quantification of erectile dysfunction drugs and some of their analogues found in counterfeit medicines. *Journal of Chromatography. A*. 1218 (37): 6439–47. doi: org/10.1016/j.chroma.2011.07.029.

Savaliya, A. A., Shah, R. P., Prasad, B., & Singh, S. (2010). Screening of Indian aphrodisiac ayurvedic/herbal healthcare products for adulteration with sildenafil, tadalafil and/or vardenafil using LC/PDA and extracted ion LC-MS/TOF. *Journal of Pharmaceutical and Biomedical Analysis*. 52 (3): 406–409. doi: org/10.1016/j.jpba.2009.05.021.

Semenistaya, E., Zvereva, I., Thomas, A., Thevis, M., Krotov, G., & Rodchenkov, G. 2015. Determination of growth hormone releasing peptides metabolites in human urine after nasal administration of GHRP-1, GHRP-2, GHRP-6, Hexarelin, and Ipamorelin. *Drug Testing and Analysis*. 7 (10): 919–925. doi: org/10.1002/dta.1787.

Sheshashena Reddy, T., Reddy, A. S., & Devi, P. S. (2006). Quantitative determination of sildenafil citrate in herbal medicinal formulations by high-performance thin-layer chromatography. *Journal of Planar Chromatography – Modern TLC*. 19 (112): 427–431. doi: org/10.1556/JPC.19.2006.6.2.

Shewiyo, D. H., Kaale, E., Risha, P. G., Dejaegher, B., Smeyers-Verbeke, J., & Van der Heyden, Y. (2009). Development and validation of a normal-phase high-performance thin layer chromatographic method for the analysis of sulfamethoxazole and trimethoprim in co-trimoxazole tablets. *Journal of Chromatography. A*. 1216 (42): 7102–7107. doi: org/10.1016/j.chroma.2009.08.076.

Shewiyo, D. H., Kaale, E., Ugullum, C., Sigonda, M. N., Risha, P. G. , Dejaegher, B., Smeyers-Verbeke, J., and Van der Heyden, Y. (2011). Development and validation of a normal-phase HPTLC method for the simultaneous analysis of lamivudine, stavudine and nevirapine in fixed-dose combination tablets. *Journal of Pharmaceutical and Biomedical Analysis* 54 (3): 445–450. doi: org/10.1016/j.jpba.2010.09.009.

Singh, S., Prasad, B., Savaliya, A., Shah, R., Gohil, V., & Kaur, A. (2009). Strategies for characterizing sildenafil, vardenafil, tadalafil and their analogues in herbal dietary supplements, and detecting counterfeit products containing these drugs. *TrAC Trends in Analytical Chemistry.* 28 (1): 13–28. doi: org/10.1016/j.trac.2008.09.002.

Skalicka-Woźniak, K., Georgiev, M. I., & Erdogan Orhan, I.. (2017). Adulteration of herbal sexual enhancers and slimmers: the wish for better sexual well-being and perfect body can be risky. *Food and Chemical Toxicology.* 108 (Pt B): 355–362. doi: org/10.1016/j.fct.2016.06.018.

Streit, R. (2017). Der Herceptin®-Fall: ein fälschungsfall von arzneimitteln größeren ausmaßes. *Bundesgesundheitsblatt – Gesundheitsforschung – Gesundheitsschutz.* 60 (11): 1203–1207. doi: org/10.1007/s00103-017-2622-2.

Stypułkowska, K., Błażewicz, A., Maurin, J., Sarna, K., & Fijałek, Z. (2011). X-ray powder diffractometry and liquid chromatography studies of sibutramine and its analogues content in herbal dietary supplements. *Journal of Pharmaceutical and Biomedical Analysis* 56 (5): 969–975. doi: org/10.1016/j.jpba.2011.08.028.

Tang, M. H. Y., Chen, S. P. L., Ng, S. W., Chan, A. Y. W., & Mak, T. W. L.. (2011). Case series on a diversity of illicit weight-reducing agents: from the well known to the unexpected. *British Journal of Clinical Pharmacology.* 71 (2): 250–253. doi: org/10.1111/j.1365-2125.2010.03822.2.

Temerdashev, A. Z., Azaryan, A. A., Labutin, A. V., Dikunets, M. A., Zvereva, I. O., Podol'skii, I. I., Berodze, G. T., & Balabaev, I. A. (2017). Application of chromatography–mass spectrometry methods to the control of sport nutrition and medicines marketed via internet. *Journal of Analytical Chemistry.* 72 (11): 1184–1192. doi: org/10.1134/S1061934817110090.

Thevis, M., Bredehöft, M., Kohler, M., & Schänzer, W. 2009. Mass spectrometry-based analysis of IGF-1 and HGH. *Doping in Sports*, edited by Thieme D. & Hemmersbach P., 195:201–7. Handbook of Experimental Pharmacology. Berlin, Heidelberg: Springer. doi: org/10.1007/978-3-540-79088-4_9.

Thevis, M., Kuuranne, T., Geyer, H., & Schänzer, W. (2012). Annual banned-substance review: analytical approaches in human sports drug testing: annual banned-substance review. *Drug Testing and Analysis.* 6 (1–2): 164–82. doi: org/10.1002/dta.1591.

Thevis, M., Thomas, A., & Schänzer, W. (2011). Doping control analysis of selected peptide hormones using LC–MS(/MS). *Forensic Science International.* 213 (1–3): 35–41. doi: org/10.1016/j.forsciint.2011.06.015.

Thomas, A., Geyer, H., Kamber, M., Schänzer, W., & Thevis, M. (2008). Mass spectrometric determination of gonadotrophin-releasing hormone (GnRH) in human urine for doping control purposes by means of LC–ESI-MS/MS. *Journal of Mass Spectrometry.* 43 (7): 908–915. doi: org/10.1002/jms.1453.

Thomas, A., Höppner, S., Geyer, H., Schänzer, W., Petrou, M., Kwiatkowska, D., Pokrywka, A., & Thevis, M. (2011). Determination of growth hormone releasing peptides (GHRP) and their major metabolites in human urine for doping controls by means of liquid chromatography mass spectrometry. *Analytical and Bioanalytical Chemistry.* 401 (2): 507–516. doi: org/10.1007/s00216-011-4702-3.

Thomas, A., Kohler M., Mester J., Geyer H., Schänzer W., Petrou M., & Thevis M. (2010). Identification of the growth-hormone-releasing peptide-2 (GHRP-2) in a nutritional supplement. *Drug Testing and Analysis*, n/a-n/a. doi.org/10.1002/dta.120.

Thomas, A., Kohler, M., Schänzer, W., Delahaut, P., & Thevis, M. 2011. Determination of IGF-1 and IGF-2, their degradation products and synthetic analogues in urine by LC-MS/MS. *The Analyst.* 136 (5): 1003–12. doi: org/10.1039/C0AN00632G.

Thomas, A., Krombholz, S., Wolf, C., & Thevis, M. (2021). Determination of ghrelin and desacyl ghrelin in human plasma and urine by means of LC–MS/MS for doping controls. *Drug Testing and Analysis.* 13 (11–12): 1862–1870. doi: org/10.1002/dta.3176.

Thomas, A., Schänzer, W., Delahaut, P., & Thevis, M. (2012). Immunoaffinity purification of peptide hormones prior to liquid chromatography–mass spectrometry in doping controls. *Methods.* 56 (2): 230–235. doi: org/10.1016/j.ymeth.2011.08.009.

Thomas, A., Walpurgis, K., Delahaut, P., Fichant, E., Schänzer, W., & Thevis, M.(2017). Determination of longR 3 -IGF-I, R 3 -IGF-I, Des1-3 IGF-I and their metabolites in human plasma samples by means of LC-MS. *Growth Hormone & IGF Research.* 35 (August): 33–39. doi: org/10.1016/j.ghir.2017.06.002.

Tie, Y., Vanhee, C., Deconinck, E., & Adams, E. (2019). Development and validation of chromatographic methods for screening and subsequent quantification of suspected illegal antimicrobial drugs encountered on the Belgian market. *Talanta* 194 (March): 876–887. doi: org/10.1016/j.talanta.2018.10.078.

Tistaert, C., Dejaegher, B., & Van der Heyden Y. 2011. Chromatographic separation techniques and data handling methods for herbal fingerprints: a review. *Analytica Chimica Acta.* 690 (2): 148–161. doi: org/10.1016/j.aca.2011.02.023.

Tomić, S., Milcić, N., Sokolić, M., & Ilić Martinac, A. (2010). Identification of counterfeit medicines for erectile dysfunction from an illegal supply chain. *Arhiv Za Higijenu Rada I Toksikologiju* 61 (1): 69–75. doi: org/10.2478/10004-1254-61-2010-1953.

Tshibangu, K. C., Worku, Z. B., De Jongh, M. A., Van Wyk, A. E., Mokwena, S. O., & Peranovic, V. (2002). Assessment of effectiveness of traditional herbal medicine in managing HIV/AIDS patients in South Africa. *East African Medical Journal* 81 (10): 499–502. doi: org/10.4314/eamj.v81i10.9231.

United States Pharmacopoeia Convention. (2021). *United Stated Pharmacopoeia (USP-NF 2021).* Frederick, USA.

Van den Broek, I., Blokland, M., Nessen, M. A., & Sterk, S. (2015). Current trends in mass spectrometry of peptides and proteins: application to veterinary and sports-doping control: mass spectrometry for detection of peptide and protein growth promoters. *Mass Spectrometry Reviews.* 34 (6): 571–592. doi: org/10.1002/mas.21419.

Vanhee, C., Francotte, A., Janvier, S., & Deconinck, E. (2020). The occurrence of putative cognitive enhancing research peptides in seized pharmaceutical preparations: an incentive for controlling agencies to prepare for future encounters of the kind. *Drug Testing and Analysis.* 12 (3): 371–381. doi: org/10.1002/dta.2717.

Vanhee, C., Janvier S., Desmedt B., Moens G., Deconinck E., De Beer J. O., & Courselle P. (2015). Analysis of illegal peptide biopharmaceuticals frequently encountered by controlling agencies. *Talanta.* 142 (September): 1–10. doi; org/10.1016/j.talanta.2015.02.022.

Vanhee, C., Janvier, S., Moens, G., Deconinck, E., and Courselle, P. (2016). A simple dilute and shoot methodology for the identification and quantification of illegal insulin. *Journal of Pharmaceutical Analysis.* 6 (5): 326–32. doi: org/10.1016/j.jpha.2016.02.006.

Vanhee, C., Moens, G., Deconinck, E., & De Beer, J. O. (2012). Identification and characterization of peptide drugs in unknown pharmaceutical preparations seized by the Belgian authorities: case report on AOD9604: identification and characterisation of peptide drugs. *Drug Testing and Analysis*. 6 (9): 964–968. doi: org/10.1002/dta.1687.

Vanhee, C., Moens, G., Van Hoeck, E., Deconinck, E., & De Beer, J. O. (2015). Identification of the small research tetra peptide Epitalon, assumed to be a potential treatment for cancer, old age and *Retinitis Pigmentosa* in two illegal pharmaceutical preparations: identification and analysis of a new illegal biopharmaceutical. *Drug Testing and Analysis*. 7 (3): 259–262. doi: org/10.1002/dta.1771.

Vanhee, C., Tuenter, E., Kamugisha, A., Canfyn, M., Moens, G., Courselle, P., Pieters, L., et al. (2018). Identification and quantification methodology for the analysis of suspected illegal dietary supplements: reference standard or no reference standard, that is the question. *Journal of Forensic Toxicology & Pharmacology*. 07 (01). doi: org/10.4172/2325-9841.1000156.

Venhuis, B. J., Vredenbregt, M. V., Kaun, N., Maurin, J. K., Fijałek, Z., & De Kaste, D.. (2011). The identification of rimonabant polymorphs, sibutramine and analogues of both in counterfeit acomplia bought on the internet. *Journal of Pharmaceutical and Biomedical Analysis*. 54 (1): 21–26. doi: org/10.1016/j.jpba.2010.07.043.

Venhuis, B. J., Keizers, P. H. J., Klausmann R., & Hegger, I. (2016). Operation resistance: a snapshot of falsified antibiotics and biopharmaceutical injectables in Europe. *Drug Testing and Analysis*. 8 (3–4): 398–401. doi: org/10.1002/dta.1888.

Venhuis, B.J. & De Kaste, D. (2012). Towards a decade of detecting new analogues of sildenafil, tadalafil and vardenafil in food supplements: a history, analytical aspects and health risks. *Journal of Pharmaceutical and Biomedical Analysis*. 69 (October): 196–208. doi: org/10.1016/j.jpba.2012.02.012.

Voss, S. C., Orie, N. N., El-Saftawy, W., Saghbazarian, S., Al-Kaabi, A., Georgakopoulos, C., Athanasiadou, I., et al. (2021). Horseradish-peroxidase-conjugated anti-erythropoietin antibodies for direct recombinant human erythropoietin detection: proof of concept. *Drug Testing and Analysis*. 13 (3): 529–538. doi: org/10.1002/dta.2957.

Walpurgis, K., Thomas, A., Lange, T., Reichel, C., Geyer, H., & Thevis M. (2018). Combined detection of the ActRII-Fc fusion proteins sotatercept (ActRIIA-Fc) and luspatercept (modified ActRIIB-Fc) in serum by means of immunoaffinity purification, tryptic digestion, and LC-MS/MS. *Drug Testing and Analysis*. 10 (11–12): 1714–1721. doi: org/10.1002/dta.2513.

Wang, J., Chen B., & Yao S. (2008). Analysis of six synthetic adulterants in herbal weight-reducing dietary supplements by LC electrospray ionization-MS. *Food Additives & Contaminants. Part A, Chemistry, Analysis, Control, Exposure & Risk Assessment*. 25 (7): 822–830. doi: org/10.1080/02652030801946553.

World Health Organisation (WHO). (2000). General guidelines for methodologies on research and evaluation of traditional medicine. Geneva, Switzerland. apps.who.int/iris/bitstream/handle/10665/66783/WHO_EDM_TRM_2000.1.pdf;jsessionid=C4FB9884E97C93CAB984CB430622F6F9?sequence=1.

———. (2018). *Substandard and Falsified Medical Products*. World Health Organisation. www.who.int/news-room/fact-sheets/detail/substandard-and-falsified-medical-products.

Wu, Y. (2006). The identification of microscopic, physical and chemistrical analysis on Curculigo orchiode and its counterfeit. *Journal of Chinese Medicinal Materials*. 29 (6): 553–552.

Yemoa, A., Habyalimana, V., Mbinze, J. K., Crickboom, V., Muhigirwa, B., Ngoya, A., Sacre, P., et al. (2017). Detection of poor quality artemisinin-based combination therapy (ACT) medicines marketed in Benin using simple and advanced analytical techniques. *Current Drug Safety.* 12 (3): 178–186. doi: org/10.2174/1574886312666170616092457.

Yu, N. H., Ho, E. N. M., Wan, T. S. M., & Wong, A. S. Y.. (2010). Doping control analysis of recombinant human erythropoietin, darbepoetin alfa and methoxy polyethylene glycol-epoetin beta in equine plasma by nano-liquid chromatography–tandem mass spectrometry. *Analytical and Bioanalytical Chemistry.* 396 (7): 2513–2521. doi: org/10.1007/s00216-010-3455-8.

Yun, J., Shin K. J., Choi J., and Jo C. (2018). Isolation and structural characterization of a novel sibutramine analogue, chlorosipentramine, in a slimming dietary supplement, by using HPLC-PDA, LC-Q-TOF/MS, FT-IR, and NMR. *Forensic Science International.* 286 (May): 199–207. doi: org/10.1016/j.forsciint.2018.03.021.

Zhu, X., Song, X., Chen. B., Zhang, F., Yao, S., Wan, Z., Yang, D., & Han, H. (2005). Simultaneous determination of sildenafil, vardenafil and tadalafil as forbidden components in natural dietary supplements for male sexual potency by high-performance liquid chromatography-electrospray ionization mass spectrometry. *Journal of Chromatography. A.* 1066 (1–2): 89–95. doi: org/10.1016/j.chroma.2005.01.038.

Zou, P., Sze-Yin Oh, S., Hou, P., Low, M., & Koh, H. (2006). Simultaneous determination of synthetic phosphodiesterase-5 inhibitors found in a dietary supplement and pre-mixed bulk powders for dietary supplements using high-performance liquid chromatography with diode array detection and liquid chromatography-electrospray ionization tandem mass spectrometry. *Journal of Chromatography. A.* 1104 (1–2): 113–122. doi: org/10.1016/j.chroma.2005.11.103.

Zvereva, I., Dudko G., & Dikunets M. (2018). Determination of GnRH and its synthetic analogues' abuse in doping control: small bioactive peptide UPLC-MS/MS method extension by addition of *in vitro* and *in vivo* metabolism data; evaluation of LH and steroid profile parameter fluctuations as suitable biomarkers. *Drug Testing and Analysis.* 10 (4): 711–722. doi: org/10.1002/dta.2256.

3 Tackling a Global Crisis
A Review of the Efficacy of LC-MS

Dong Hyeon Shin[1] and Ronny Priefer[1,]*
[1]MCPHS University, 179 Longwood Ave, Boston, MA, USA, 02115

CONTENTS

3.1 Introduction 77
3.2 Chromatography Coupled with Mass Spectrometry 78
3.3 Counterfeit Drugs for Pathogenic Diseases...79
3.4 Counterfeit Drugs for Erectile Dysfunction 82
3.5 Counterfeit Drugs for Steroids...85
3.6 Counterfeit Drugs for Obesity...86
3.7 Concerning Counterfeit Drugs with Detrimental Effects...........................88
3.8 Conclusion...90
References..91

3.1 INTRODUCTION

People in low and middle-income countries are prone to consume low-quality medicines that can cause deleterious effects on the patient and the population as a whole. According to the Centers for Disease Control and Prevention (CDC), 9%–41% of medicines sold in these countries are counterfeit (1). Counterfeit medicines are classified as either poor-quality, having sub-standard levels, or containing no active ingredient/s. As a result, there are increased risks of treatment failure, prolonged illness, adverse reactions, and higher levels of morbidity and mortality (2,3).

Based on the World Health Organization's (WHO) estimates, over one million deaths occur annually as a result of counterfeit drugs (4). In developed countries, governments and organizations have put significant efforts toward combatting their distribution and production (5). However, due to lack of either infrastructure or funds, this is a limitation in developing countries, thus leading to an increase in the sales of these counterfeit medicines. It is estimated that more than 50% drugs sold in parts of Africa and Asia are counterfeit (6). Counterfeit drugs are generally less expensive, making them more affordable to the population living in low-income countries. Moreover, those living in developing countries suffer from diseases that

DOI: 10.1201/9781003270461-3

are fairly controlled in developed countries. For example, a study on the quality of antibiotics in Ghana, noted that over 66% of the sampled antibiotic products were substandard (7). The issue of counterfeit drugs is further exasperated due to governments and pharmaceutical companies in developing countries' hesitation to report the prevalence of the issue (6). It has been suggested that this is due to a fear that the public will lose faith in the healthcare system, in addition to pharmaceutical companies earning less profit (6,8). Overall, the prevalence of counterfeit drugs has a detrimental effect on healthcare and risks billions of people's lives, especially those in developing countries. Fortunately, various technologies are being used to help mitigate this worldwide concern.

Liquid chromatography-mass spectrometry (LC-MS) is a powerful technology for determining and analyzing ingredients within a drug formulation. Due to its sensitivity and specificity, LC-MS is extensively used in therapeutic drug monitoring (9,10,11), clinical forensic toxicology (12,13,14,15,16), detection of drugs in serum and urine (17,18,19,20,21,22), and screening for drugs of abuse (22,23,24). Additionally, LC-MS has been employed to detect, identify, and quantify drug metabolites (18,25,26,27). For example, it was possible to differentiate the metabolites of aspirin, arising from enteric-coated versus simple formulation, with the use of LC-MS (17). Immunoassays are the most common technique used in therapeutic drug monitoring; however, this technique has limitations for certain drugs and may show lack of accuracy (9). LC-MS can reduce the time of analysis, providing a detailed pharmacokinetic profile (9).

LC-MS has also shown a variety of uses and advantages in healthcare by identifying and detecting compounds within pharmaceutical products. Herein, is a review of the efficacy and application of LC-MS in terms of helping to alleviate the global counterfeit drug issue.

3.2   CHROMATOGRAPHY COUPLED WITH MASS SPECTROMETRY

Chromatography is an analytical technique involving the physical separation of chemical compounds (28). Because of its accuracy and effectiveness, chromatography has become widely accepted and is used in several applications (29). High-performance liquid chromatography (HPLC) is one of most utilized methods, with many labs favoring it due to its high sensitivity and specificity (30). In HPLC, a desired analyte or component can be separated from a mixture based on its affinity toward either a mobile or stationary phase. Depending on the type of solvent and column employed, the retention time (RT) can be varied. Hydrophobic compounds have a strong affinity for the nonpolar stationary phase, thus have a longer RT (8). Conversely, hydrophilic molecules will travel more quickly through the column when utilizing a polar stationary phase. The polarity index of the solvent helps in choosing the optimal eluent in liquid chromatography (LC).

LC-MS is the coupling of the physical separation, via chromatography, followed by detection from the measurement of the mass of the compound through mass spectrometry (MS) (28). LC-MS provides advantages of detecting a diversity of compounds, such as drugs, extract products, adulterated foods, pesticides, and so forth (28). The MS works by first ionizing a compound so that it can be separated from similarly weighted compounds, then detected, based on their mass-to-charge ratio (m/z) (31). There are two main ion sources employed: electrospray ionization (ESI) and atmospheric pressure ionization. MS provides additional improvements on identification and quantification of compounds by analyzing the fragmentation pattern (i.e., daughter peaks) to obtain a "fingerprint" of the compound (8). However, there are several limitations that prevent developing countries from routinely using this technology. Specifically, LC-MS is expensive, has low sensitivity for certain specific compounds, needs a high level of training, and requires a high-power source to operate (8).

3.3 COUNTERFEIT DRUGS FOR PATHOGENIC DISEASES

Antibiotics, antimalarials, and antiparasitics are drug classes that are often counterfeited or substandard in developing countries (32). Sub-therapeutic anti-infective drugs not only harm an individual's health, but also increases the risk of drug resistance (32). Thus, it is crucial to have a method of identifying these poor-quality, anti-infective drugs (Table 3.1).

Populations in Ghana, a low-income country, face issues of counterfeit medications. It was noted that in Ghana there was a high percentage of antibiotic drugs being of poor quality (7). As a result of this, there has been an emergence of resistant bacterial strains (7). In one study, the quality of antibiotics were assessed based on 348 sampled products from both authorized sales outlets (AT) (i.e., hospitals, pharmacies, and licensed chemical shops) and unauthorized sales outlets (UAT), such as street sales, in Ghana (7). They evaluated the most commonly used antibiotics, specifically: amoxicillin, amoxicillin/clavulanic acid, ampicillin, metronidazole, sulphamethoxazole/trimethoprim, ceftriaxone, cefuroxime, tetracycline, ciprofloxacin, benzylpenicillin, phenoxymethylpenicillin, erythromycin, and gentamicin (7). LC coupled to tandem MS (LC-MS/MS) was used to determine the contents of these antibiotics.

Based on the British Pharmacopoeial standard, the drugs' quality were evaluated in terms of percentage of active pharmaceutical ingredient (API) (7). Only 117 (33.6%) of the 348 sampled products met the standard of having between 90% and 110% of the API (7). Conversely, 66% of the sampled products had sub-therapeutic amounts of the API, having either 90% or 110%. Approximately 60% of samples from AT were substandard, while an astonishing 90% of those from UAT were counterfeit (7). Although, LC-MS/MS was beneficial to assess the quality of the antibiotics from AT and UAT in Ghana, it did not alleviate the problem.

A similar study evaluated several drug products, including tablets, capsules, suspensions, syrups, injection solutions, and ear and eye drops, of 13 antibiotics used in Ghana (33). The API analyzed were: penicillin G, tetracycline, erythromycin, ceftriaxone, gentamicin, cefuroxime, ampicillin, metronidazole, ciprofloxacin, clavulanic acid, trimethoprim and sulphamethoxazole, and amoxicillin (33). The drug products were collected from government hospitals, privately owned pharmacies, licensed chemical shops, and informal suppliers throughout the country (33). LC-MS/MS was used to screen for the antimicrobial quality of these products and was characterized based on European medicines agency guidelines, in terms of percentage of API (33). Among a total of 20 samples (10 tablets and 10 capsules), erythromycin had the lowest percentage of API at <41% (33). In addition, only samples of four antibiotics (i.e., amoxicillin, clavulanic acid, ampicillin, trimethoprim) were >90% of API (#33). In the sampled suspensions and syrups, only three antibiotics (i.e., sulphamethoxazole, ciprofloxacin, and erythromycin) were >90%, while ten were <18% (33). By using LC-MS/MS, it is found that intravenous samples had highest percentage of quality, while tablets and capsules had the lowest (33). Overall, the study revealed that these variations in percent API among various drug products puts extreme risk on the healthcare system in Ghana.

Malaria is a major health problem in Southeast Asia, with an estimated eight million cases and 11,600 deaths in 2018 (34). Comparatively, only 2,000 cases of malaria are observed in United States every year (35). Unless properly treated, this parasitic disease can be severely detrimental and fatal. Due to its prevalence in Southeast Asia, antimalaria drugs make up a large portion of pharmaceutical sales (36). A recent study revealed that 38% of samples of the antimalaria drug, artesunate, were counterfeit (36). In addition, the study suggests that the active drug may have actually been substituted with chloroquine (36).

Halfan, is an antimalaria agent commonly used in Southeast Asia. The active ingredient found within, is halofantrine hydrochloride (36). There has been suspicion that large amounts of this product are counterfeit, thus leading to treatment failure. It was noted that determining whether a product contains "wrong" or "unknown" active ingredient is a challenge. To solve for this complexity, LC-MS (and other analytical techniques) was employed to identify suspected counterfeits (36,37,38,39,40).

LC-MS was run on suspected counterfeit Halfan suspensions, ultimately leading to the identification of sulfamethazine, an antibacterial drug (36). In addition, other compounds were present within the suspension that were not halofantrine hydrochloride. It was concluded that the counterfeit Halfan was substituted with an antibacterial agent, thus not an antimalaria, which places high risk of treatment failure (36).

Another study in Southeast Asia analyzed 50 mg tablet of artesunate by using LC-MS (41). Artesunate is an antimalarial agent that is usually used in patients infected with drug-resistant malaria (42). Similar to the aforementioned concerns with Halfan suspension, there has been prevalence of counterfeit artesunate. It was speculated that numerous artesunate tablets contained sub-therapeutic amounts, which may lead to treatment failure.

TABLE 3.1
Reported examples of counterfeit medications for pathogenic diseases by using LC-MS

Source	Preparation types	Study aim	Counterfeit	Reference
Ghana	Tablets, capsules, syrups, suspensions, injectables, ear and eye drops	Assessed the quality of antibiotics in terms of API	66% of 348 samples had sub-therapeutic amounts of the API	#7
Ghana	Tablets, capsules, suspensions, syrups, injection solutions, ear and eye drops	Analyzed 13 antibiotics in terms of percentage of API	Tablets: - Erythromycin samples <41% of API Suspensions and syrups: - ten antibiotics (penicillin G, tetracycline, ceftriaxone, gentamicin, cefuroxime, ampicillin, metronidazole, clavulanic acid, trimethoprim, and amoxicillin) <18% of API	#33
Southeast Asia	Suspensions	Determined a counterfeit Halfan suspensions	Sulfamethazine (anti-bacterial)	#36
Southeast Asia	Tablets	Determined counterfeit 50 mg tablets of artesunate	23 (68%) of the 34 artesunate products did not contain the expected active ingredient at all	#41
DRC	Tablets, suspensions	Assessed the quality of AL formulations	- 46 (30.7%) of the 150 samples had sub-therapeutic amounts of artemether - 32 (21.7%) samples had sub-therapeutic amounts of lumefantrine	#42
Bangladesh	Capsules	Identified counterfeit 'Miltefos' capsules	The capsules contained two excipients, without any API	#32

A variety of 50 mg tablets were collected from Vietnam, Burma, Thailand, Laos, and Cambodia (41). Using LC-MS, it was determined that some tablets from Cambodia contained only 21 mg of artesunate, thus sub-therapeutic (41). More telling however, was that 23 (68%) of the 34 artesunate products, did not contain the expected active ingredient at all (41). Through LC-MS, it was determined that some of the counterfeit artesunate products actually contained erythromycin A, which is an antibacterial and not an antimalaria medicine (41).

Similar to Southeast Asia, malaria incidences and death rates in the Democratic Republic of Congo (DRC) are high. The disease was responsible for 36% of the total deaths in 2014, and roughly 11.6 million malaria cases were reported in 2015 (43). Artemisinin-based combination therapy (ACT) is commonly used for treating *Plasmodium falciparum* malaria in DRC (43). ACT is composed of either artesunate, artemether, or dihydroartemisinin, plus one additional antimalarial agent, such as amodiaquine or lumefantrine (43). Unfortunately, there has been an increase in reports of poor-quality or substandard pharmaceutical ACT products.

A study utilizing LC-MS and other analytical techniques, assessed the quality of antimalarial artemether-lumefantrine (AL) formulations in the DRC (37,38,41,43). A total of 150 AL samples were collected from private pharmaceutical outlets, which are the main source of sales within the DRC (43). Based on LC-MS, 46 (30.7%) of the 150 samples had amounts of artemether <90% (sub-therapeutic), while 17 (11.3%) had >110%, which exceeds the amount claimed (43). With respect to lumefantrine, 32 (21.7%) samples had <90%, while 8 (5.3%) samples were >110% (43).

Black fever (also known as kala-azar) is a parasitic disease caused by the protozoan, *Visceral leishmaniasis,* and is extremely fatal if not treated properly (44). Miltefosine, is a first-line, orally available, medication indicated to treat black fever, (32) and is widely used in Bangladesh. However, in Bangladesh, there has been a noticeable emergence of treatment failure with "Miltefos" (32). Miltefos should contain the API miltefosine, however studies have suggested counterfeit issues.

Miltefos was evaluated by assessing its pharmaceutical equivalence to Impavido, another pharmaceutical product also containing miltefosin (32). LC-MS/MS was utilized to determine if the Miltefos capsules contained miltefosine. Astonishingly, it was found that the capsules only contained two excipients, without any actual API (32).

3.4 COUNTERFEIT DRUGS FOR ERECTILE DYSFUNCTION

In addition to healthcare concerns with infectious diseases, there has been a recognizable concern in patients treating for erectile dysfunction (ED). It is estimated that the global prevalence of ED is in the range of 3%–76.5% (45), and affects 30 million men in the US alone (46). It is predicted that by 2025, ED will increase to 322 million men worldwide(47).

Phosphodiesterase type-5 inhibitors (PDE-5Is) are widely used to treat ED. With the high prevalence of the disease, there has been a noticeable growth in sales and distribution of counterfeit and illicit PDE-5Is (48). These products are generally marketed as "sexual enhancer drugs," which are easily accessible without a prescription. Also, numerous dietary supplements or herbal medicines are labeled to improve sexual functional. However, these products may have been adulterated with PDE-5Is, such as sildenafil, tadalafil, vardenafil, and other analogs (49) (Table 3.2). Without the proper prescription, potential harm and risks are associated with these counterfeit PDE-5Is. For example, if patients take a counterfeit product containing nitroglycerin, they may experience severe low blood pressure (48).

One study screened for products marketed as sexual enhancers and analyzed compounds within, by utilizing LC-MS in addition to other analytical techniques. A total of 181 counterfeit products were collected from various online sites, direct purchases, international post services, and offline markets (48). Samples were in forms of tablets, capsules, pills, powders, liquids, and films (48). It was determined that a staggering 86.2% of all samples contained PDE-5Is and/or their analogues (48). Furthermore, 49.4% of all samples contained two or more PDE-5Is (48). The study concluded by emphasizing that mixtures of these components may cause severe toxicities (48).

A similar study assessed products that claimed to be effective in improving sexual performance. These included foods, dietary supplements, drugs, and herbal medicines (50). A total of 362 samples were collected from the online or offline markets between 2014 and 2017 (50). By using LC-MS/MS, the study found that 40% of all samples contained one or more PDE-5Is (50). Among the samples, food had the highest detection of PDE-5Is (51%) followed by counterfeit drugs (32%), dietary supplements (12%), and herbal medicines (6%) (50). In addition, sildenafil (50%) was the most commonly found API in these adulterated products, followed by tadalafil (41%) (50).

Another study analyzed for PDE-5Is in dietary supplements by using LC-MS/MS. A total of 13 samples (five tablets and eight capsules) were collected from internet websites (51). Samples were labeled as containing an ingredient from herbal plants, such as butea superb root, gingko, maritime pine bark, ginseng root, reishi fruit, and so forth (51). One of 13 samples was found to be counterfeit, containing high levels of tadalafil, in addition to low levels of sildenafil (51).

The US FDA Forensic Chemistry Center evaluated 40 samples of traditional herbal medicines (THM) marketed as sexual enhancers by using LC-MS/MS. The study revealed that 19 (47.5%) of the samples were counterfeit and contained either sildenafil, tadalafil, and/or homosildenafil (52). Additionally, it was seen that these products were not only adulterated with PDE-5Is, but also with an analogue, homosildenafil (52). Finally, it was found that the capsule shells, used as the vehicle, were adulterated to hide the drug substances (52).

Many additional studies have analyzed and screened for PDE-5Is in dietary supplements, herbal products, and sexual enhancers. One aimed to identify and

TABLE 3.2
Reported examples of counterfeit medications for erectile dysfunction (ED) by using LC-MS

Source	Preparation types	Study aim	Counterfeit	Reference
Various	Tablets, capsules, pills, powders, liquids, films	Screened for products marketed as sexual enhancers and analyzed compounds within	86.2% of 181 samples contained PDE-5Is and/or their analogues	#48
South Korea	Foods, dietary supplements, drugs, herbal medicines	Assessed for products claimed as sexual enhancers	40% of 362 samples contained one or more PDE-5Is	#50
Various	Tablets, capsules	Analyzed for PDE-5Is in dietary supplements	1 (7.7%) of 13 samples was found to be counterfeit	#51
Various	Tablets, capsules, bulk powders	Evaluated traditional herbal medicines marketed as sexual enhancers	19 (47.5%) of 40 samples were found to be counterfeit and contained either sildenafil, tadalafil, and/or homosildenafil	#52
U.S	Tablets	Identified API in FDA approved Cialis (tadalafil) tablets	4 tablets contained sildenafil	#53
U.S	Tablets	Screened a single blister pack containing 100 mg tablets of Viagra	1 (50%) of 2 tablets contained no detectable amounts of Viagra	#54

quantify any API in FDA approved Cialis (tadalafil) tablets. Four suspected blisters of Cialis tablets were analyzed using LC coupled to a diode array detection tandem MS (LC-DAD-MS) (53). All samples were found to contain sildenafil, which is not the API of Cialis. One sample contained a therapeutic relevant dose of sildenafil, while the other three samples had sub-therapeutic quantities (53). Tadalafil was only found in three samples, all of which also contained sub-therapeutic amounts of sildenafil (53). In addition, one sample contained a mixture of tadalafil and *trans*-tadalafil (stereoisomer of tadalafil), which should not be present in Cialis tablets

(53). Overall, all samples were recognized as counterfeit by either containing sildenafil or a combination of sildenafil and tadalafil.

Another study screened a single blister pack labelled to contain 100 mg tablets of Viagra, by using HPLC-DAD-MS/MS (54). The investigators analyzed two tablets, and found that one contained 98.4 mg of the correct API, however, the second had no detectable amounts of Viagra (4). The authors concluded that dose-to-dose variability among the tablets is a significant concern.

3.5   COUNTERFEIT DRUGS FOR STEROIDS

Similar to the issue of PDE-5Is, anabolic-androgenic steroids (AAS) are illegally used to a great extent, and a large number of drugs are counterfeit (Table 3.3). Clinically, AAS, such as testosterone, are used as androgen replacement therapy (ART) and require a prescription (55). ART is used in men who have low levels of testosterone and for treatment of hypogonadism (56). ART can help restore physiological testosterone levels to prevent complications, such as osteoporosis and sarcopenia (57). However, many athletes have abused AAS for the benefits of muscle growth, strength gain, and performance enhancement. Currently, there is concern that certain dietary supplements may contain AAS, with these products also being easily accessible to consumers. It was noted that these products may pose a significant risk of adverse effects. Many analytical techniques, including LC-MS, have been used to detect for these adulterated products.

One study evaluated AAS in products by using ultra-liquid chromatography-tandem mass spectrometry (UHPLC-MS/MS). A total of 19 samples from the internet and offline markets were analyzed (58). Nine (50%) of the samples contained AAS with some having two or more of these present (58). Testosterone and testosterone 17-propionate (26%) were the highest percentage of detected AAS compounds, followed by boldenone (21%) and testosterone 17-valerate (16%) (58).

In addition to the problem with AAS, corticosteroids are sometimes counterfeited in various products (Table 3.3). Corticosteroids are extensively used for their anti-inflammatory effects, which are beneficial in a variety of diseases, such as psoriasis, eczema, arthritis, asthma, rhinitis, inflammatory bowel disease, and so forth. (59). Due to their wide range of uses, herbal medicines and cosmetic products are marketed and advertised to treat these conditions. However, these products may cause potential side effects, such as hypertension, diabetes, osteoporosis, Cushing's syndrome, and others. (59). Due to concerns of these undesirable adverse reactions, LC-MS was employed to screen for counterfeit products.

A study utilized LC-ESI-MS to screen for corticosteroids in a total of 11 cosmetic products, formulated as either creams (2), powders (6), or capsules (3), which were purchased from local Italian markets (59). It was revealed that one cream did indeed contain betamethasone 17-valerate and betamethasone 21-valerate (59).

TABLE 3.3
Reported examples of counterfeit medications for steroids by using LC-MS

Source	Preparation types	Study aim	Counterfeit	Reference
Various	Tablets, liquid injectables	Evaluated products that are advertised to increase muscle mass	9 (50%) of 19 samples contained AAS	#58
Italia	Creams, powders, capsules	Screened for corticosteroids in cosmetic products	1 (9%) of 11 samples contained corticosteroids	#59
India	Tablets, capsules, powders	Screened for corticosteroids in ayurvedic/ herbal healthcare products	11 (18.9%) of 58 samples contained corticosteroid	#60

Another larger study detected synthetic corticosteroids in ayurvedic/herbal healthcare products (AHP) by using LC-MS/time of flight (LC-MS/TOF). A total of 58 AHPs were collected from regions of India at chemist shops (45) or practitioners of alternative medicines (13) (60). Eleven (18.9%) of the samples contained the corticosteroid, dexamethasone (60). Interestingly, all eleven of these counterfeit samples were dispensed by practitioners of alternative medicines (60).

3.6 COUNTERFEIT DRUGS FOR OBESITY

Obesity is a major global chronic disease, with a growing trend. Since 1975, it is estimated that global obesity has nearly tripled (61). In 2016, 650 million adults, aged 18 years and older, were classified as obese, accounting for ~13% of the world's adult population (61). In addition, over 340 million children and adolescents aged 5–19 were classified as obese in 2016 (61). Sadly, in 2020, it was estimated that 39 million children *under* the age of five were obese (61).

Major health concerns with regards to obesity, are related to its complications, predominately heart disease, stroke, and Type 2 diabetes (62). These complications account for billions of dollars of annual medical costs. In the US alone, this led to nearly $150 billion per year (62). Recently, numerous dietary supplements and herbal products are marketed and sold as weight loss supplements, which had also given rise to a counterfeit market.

TABLE 3.4

Reported examples of counterfeit medications for obesity by using LC-MS

Source	Preparation types	Study aim	Counterfeit	Reference
Various	Tablets, capsules	Screened for sibutramine in counterfeit dietary supplements	11 (64.7%) of 17 samples were positive for sibutramine	#39
Various	Capsules, sachets	Monitored for sibutramine and its analogue in instant coffee brand and a herbal weight loss product	- 8 (44%) out of 18 sachets of the coffee contained DDMS - All 5 capsules contained sibutramine	#54
China	Capsules, teabags	Analyzed for sibutramine and its analogues in weight reducing dietary herbal supplements	11 (50%) of 22 samples were positive for sibutramine and N-mono-desmethyl-sibutramine	#41

Sibutramine is an anti-obesity drug that was approved by the FDA in 1997, but withdrawn in 2010 due to its potential carcinogenicity and other adverse reactions (39). Unfortunately, sibutramine has been adulterated and counterfeited in numerous dietary supplements and herbal products (Table 3.4).

A study utilized LC-MS/MS to screen and detect for sibutramine and its analogues in counterfeit dietary supplements. The researchers obtained 17 samples from online websites (5) and US Customs Ports of Entry (12) that were labeled as containing derivatives from plant seeds, leaves, herb, or tea (39). In total, 11 (64.7%) of the samples were positive for synthetic sibutramine. Furthermore, these samples contained low levels of N-desmethyl-sibutramine (DMS), an analogue of sibutramine (39).

Another study monitored for sibutramine in an instant coffee brand and herbal weight loss product using LC-DAD-MS/MS. They screened a total of 18 sachets of the coffee and five capsules of the herbal product (54). With the coffee, 8 (44%) sachets contained didesmethyl-sibutramine (DDMS), an analogue of sibutramine, while the remaining 10 contained no API (54). In the analysis of the herbal weight loss products, all five (100%) capsules contained sibutramine, DDMS, DMS, and benzyl-sibutramine, the latter of which is a rarely reported analogue of sibutramine. The dose range of sibutramine in the capsules were 23–30 mg, which is 2–5 times higher than the previously commercially available Reductil or Meridia (54). Additionally, the benzyl-sibutramine made up 9%–13% of the total sibutramine content (54).

A similar study analyzed for sibutramine and its analogues in weight reducing dietary herbal supplements by employing LC-ESI-MS. Twenty-two samples of teabags and capsules, that were labeled as containing only natural supplements, without any synthetic drugs were evaluated (41). However, a staggering 11 (50%) of these were positive for sibutramine and N-mono-desmethyl-sibutramine (41).

3.7   CONCERNING COUNTERFEIT DRUGS WITH DETRIMENTAL EFFECTS

Similar to the issues related to ED, steroids, and obesity, other counterfeit drugs, such as abortion-inducing drugs, opioids, and anti-diabetic drugs, are becoming problematic (Table 3.5). Commonly, these medications require medical supervision to prevent detrimental adverse reactions (63,64,65). However, these drugs are falsely marketed and distributed through illegal markets, especially online stores (63,64,65). Several studies highlighted the importance of LC-MS in detecting these adulterants.

Abortion displays complex of moral, ethical, and social issues, and it requires multiple processes for it to be done safely. Consequently, more individuals are seeking illegal abortion-inducing drugs due to ease accessibility (63). Globally, there were approximately 55.7 million abortions between 2010 to 2014, of which 45% deemed as unsafe (66). In addition, according to the Korean Ministry of Food and Drug Safety, counterfeit abortion-inducing drugs have increased tenfold in 2018 compared to 2016 (63). It is noted that these drugs may cause various side effects, such as uterine rupture, irregular heart, hypertension, hypotension, anemia, pulmonary edema, and even death (63). A study utilized LC-MS/MS to analyzed for abortion-inducing compounds in counterfeit drugs. Two tablets were collected from an illegal market in Korea that were advertised for abortion-inducing function, without having a proper label on the package (63). One tablet contained mifepristone, an active abortion-inducing drug, while the other also possessed misoprostol, another abortion causing drug (63). The concentrations of detected compounds in the tablets were higher than that of Mifegyne, an oral contraception widely used in the US and Europe, which increases patient risk (63). The authors noted that, despite increased use of abortion-inducing counterfeit medications, there is not enough studies regarding this problem and thus should be more regularly screened to protect public health (63).

Similar to the abortion-inducing drugs, opioids have harmful side effects when misused. Tramadol, an opioid, is generally used as analgesic in patients with moderate to severe pain. However, tramadol can be abused to treat premature ejaculation, anxiety, and depression (64). There has been a high emergence of branded products containing illicit amounts of tramadol that lead to lethal intoxication or seizure (64). In addition, some products are co-formulated with other central nervous system depressants, such as benzodiazepines (64). It is concerning that these combination products may exasperate unfavorable adverse reactions. One study analyzed for tramadol and some coformulated compounds in counterfeit drugs by using LC-MS/MS. Ten tablets of tramadol from Egyptian markets, which were branded as Tramadol-X, Tamol-X, Supper tramadol-X, Tee-doll, and Tramajack, were collected (64). Among these samples, Tramadol-X and Supper tramadol-X were counterfeit, containing different doses of tramadol compared to genuine tramadol tablets (64). Additionally, it was discovered that additional drugs, such as alprazolam, diazepam, chlorpheniramine maleate, diphenylhidramine, and paracetamol, were coformulated with tramadol (64).

Regarding anti-diabetics, these vital drugs are well known for having several side effects. Due to this, numerous patients have explored traditional Chinese herbal medicines as an alternative (65). However, there has been increasing reports of adulteration with "natural" herbal products that are marketed as anti-diabetic (65). Some patients who consumed these reported experiencing adverse reactions, such as gastrointestinal symptoms, hypoglycemia, skin allergies, and other side effects, that are typically associated with anti-diabetics (65). A study screened for ten synthetic hypoglycemic drugs in "natural" anti-diabetic herbal products by utilizing LC-MS/MS. These ten drugs were: gliquidone, glipizide, gliclazide, glibenclamide, glimepiride, rosiglitazone, repaglinide, metformin, phenformin, and tolbutamide (65). A total of 20 samples were collected from Chinese markets. Thirteen (65%) of these were adulterated with synthetic hypoglycemic drugs (65). Glibenclamide (5 samples) was found to be the most common adulterant, followed by gliclazide and phenformin (3 each), then glimepiride and metformin (1 each) (65). Surprisingly, in some samples, the capsule shells themselves were adulterated to hinder detection from a regular analytical processes (65).

Unfortunately, in addition to small molecules, biologics have seen an increase in counterfeit medications (Table 3.5). Copious cases reports highlight several protein and peptide drugs that are illegally manufactured and are available to the public prior to completing clinical trials (67). It has been noted that these counterfeit biological drugs may have severe healthcare ramifications (67). LC-MS is one of the most powerful analytical techniques for detecting these macromolecules.

The Belgium Federal Agency for Medicinal and Health Products employed LC-MS/MS to detect for counterfeit injectable peptides (67). The study analyzed a collection of 65 samples taken between 2009 and 2014 (67). Among these samples, the majority were in unlabeled vials (56). The biologics that were screened for were: doping peptides (AOD9604, CJC1295, sermorelin, GHRP-2, GHRP-6, Ipamorelin, Thymosin beta 4), a skin tanning peptide (Melanotan II), and a potential geroprotective agent (Epitalon) (67). Thirty-three (50.7%) of the samples were positive for a screened peptides, of which 19 were from unlabeled vials (67). In addition, 31 of the positive samples contained one screened peptide while the other two had a mixture of two peptides (67). Furthermore, doping peptides (63.6%) were found to be the most abundant among the positive samples, followed by Melanotan II (30.3%) and Epitalon (6.1%) (67). LC-MS is one of the few analytical techniques available to screen for these biologics.

TABLE 3.5
Reported examples of counterfeit medications for concerning drugs

Source	Preparation types	Study aim	Counterfeit	Reference
South Korea	Tablets	Analyzed for abortion-inducing compounds in counterfeit drugs	One tablet contained mifepristone, while the other also had misoprostol	#63
Egypt	Tablets	Analyzed for tramadol and some co-formulated compounds in counterfeit products	- Samples from Tramadol-X and Supper tramadol-X were counterfeit - Some other drugs, such as alprazolam, diazepam, chlorpheniramine maleate, diphenylhidramine, and paracetamol, were co-formulated with tramadol	#64
China	Tablets, capsules	Evaluated ten synthetic hypoglycemic drugs in 'natural' anti-diabetic herbal products	13 (65%) of 20 samples were adulterated with synthetic hypoglycemic drugs	#65
Various	Injectable solution	Detected for counterfeit injectable peptides	33 (50.7%) of 65 samples were positive for screened peptides	#67

3.8   CONCLUSION

Counterfeit drugs are applied to products that are falsified or mislabeled. These products may include either having wrong ingredients, lack or insufficient amounts of API, or be in fake packaging (68). Counterfeit drugs are a global concern, threatening the entire healthcare system. Antibiotics, antimalarials, and antiparasitics are commonly counterfeit in developing countries due to the prevalence of infectious diseases. In developed countries, dietary supplements and herbal medicines are commonly adulterated and marketed to target various conditions, such as ED, obesity, diabetes, and others. Moreover, adulterated products are widely distributed through online markets, making these easily accessible. Biological pharmaceutical products are also seeing an increase as counterfeit. Detection of any of these is crucial to mitigate the risk associated with their misuse. In an attempt to understand the scope of the problem, LC-MS has been extensively utilized. Numerous studies validate the efficacy and advantages of LC-MS in identifying these counterfeit drugs.

REFERENCES

1. Centers for Disease Control and Prevention. *Counterfeit medicines*. Centers for Disease Control and Prevention. Retrieved March 24, 2022, from wwwnc.cdc.gov/travel/page/counterfeit-medicine
2. Ozawa, S., Evans, D. R., Bessias, S., Haynie, D. G., Yemeke, T. T., Laing, S. K., & Herrington, J. E. (2018). Prevalence and estimated economic burden of substandard and falsified medicines in low- and middle-income countries. *JAMA Network Open, 1*(4), 1–22. doi.org/10.1001/jamanetworkopen.2018.1662
3. Mackey, T. K., & Liang, B. A. (2011). The global counterfeit drug trade: patient safety and public health risks. *Journal of Pharmaceutical Sciences, 100*(11), 4571–4579. doi.org/10.1002/jps.22679
4. Orton, T. (2019, May 1). *A global model to reduce deaths from counterfeit drugs in the pharmaceutical industry*. American University. Retrieved March 25, 2022, from www.american.edu/kogod/news/a-global-model-to-reduce-deaths-from-counterfeit-drugs-in-the-pharmaceutical-industry.cfm
5. OECD iLibrary. *Trade in counterfeit pharmaceutical products*. Retrieved March 25, 2022, from www.oecd-ilibrary.org/governance/trade-in-counterfeit-pharmaceutical-products_a7c7e054-en
6. Cockburn, R., Newton, P. N., Agyarko, E. K., Akunyili, D., & White, N. J. (2007). The global threat of counterfeit drugs: why industry and governments must communicate the dangers. *PLoS Medicine, 2*(4), 302–308. doi.org/10.1371/journal.pmed.0020100
7. Bekoe, S. O., Ahiabu, M. A., Orman, E., Tersbøl, B. P., Adosraku, R. K., Hansen, M., Frimodt-Moller, N., & Styrishave, B. (2020). Exposure of consumers to substandard antibiotics from selected authorised and unauthorised medicine sales outlets in Ghana. *Tropical Medicine & International Health, 25*(8), 962–975. doi.org/10.1111/tmi.13442
8. Bolla, A. S., Patel, A. R., & Priefer, R. (2020). The silent development of counterfeit medications in developing countries – a systematic review of detection technologies. *International Journal of Pharmaceutics, 587*, 119702. doi.org/10.1016/j.ijpharm.2020.119702
9. Saint-Marcoux, F., Sauvage, F.-L., & Marquet, P. (2007). Current role of LC-MS in therapeutic drug monitoring. *Analytical and Bioanalytical Chemistry, 388*(7), 1327–1349. doi.org/10.1007/s00216-007-1320-1
10. Decosterd, L. A., Widmer, N., André, P., Aouri, M., & Buclin, T. (2016). The emerging role of multiplex tandem mass spectrometry analysis for therapeutic drug monitoring and personalized medicine. *TrAC Trends in Analytical Chemistry, 84*, 5–13. doi.org/10.1016/j.trac.2016.03.019

11. Taylor, P. J., Tai, C.-H., Franklin, M. E., & Pillans, P. I. (2011). The current role of liquid chromatography-tandem mass spectrometry in therapeutic drug monitoring of immunosuppressant and antiretroviral drugs. *Clinical Biochemistry*, *44*(1), 14–20. doi.org/10.1016/j.clinbiochem.2010.06.012

12. Maurer, H. H. (2007). Current role of liquid chromatography–mass spectrometry in clinical and forensic toxicology. *Analytical and Bioanalytical Chemistry*, *388*(7), 1315–1325. doi.org/10.1007/s00216-007-1248-5

13. Rentsch, K. M. (2016). Knowing the unknown – state of the art of LCMS in toxicology. *TrAC Trends in Analytical Chemistry*, *84*, 88–93. doi.org/10.1016/j.trac.2016.01.028

14. Deventer, K., Pozo, O. J., Verstraete, A. G., & Van Eenoo, P. (2014). Dilute-and-shoot-liquid chromatography-mass spectrometry for urine analysis in doping control and analytical toxicology. *TrAC Trends in Analytical Chemistry*, *55*, 1–13. doi.org/10.1016/j.trac.2013.10.012

15. Maurer, H. H. (2004). Position of chromatographic techniques in screening for detection of drugs or poisons in clinical and forensic toxicology and/or doping control. *Clinical Chemistry and Laboratory Medicine (CCLM)*, *42*(11), 1310–1324. doi.org/10.1515/cclm.2004.250

16. Couchman, L., & Morgan, P. E. (2010). LC-MS in analytical toxicology: Some practical considerations. *Biomedical Chromatography*, *25*(1-2), 100–123. doi.org/10.1002/bmc.1566

17. Dei Cas, M., Rizzo, J., Scavone, M., Femia, E., Podda, G. M., Bossi, E., Bignotto, M., Caberlon, S., Cattaneo, M., & Paroni, R. (2021). In-vitro and in-vivo metabolism of different aspirin formulations studied by a validated liquid chromatography tandem mass spectrometry method. *Scientific Reports*, *11*(1), 1–10. doi.org/10.1038/s41598-021-89671-w

18. Liu, Y., Li, Y., Zhang, T., Zhao, H., Fan, S., Cai, X., Liu, Y., et al. (2020). Analysis of biomarkers and metabolic pathways in patients with unstable angina based on ultra-high-performance liquid chromatography-quadrupole time-of-flight mass spectrometry. *Molecular Medicine Reports*, *22*(5), 3862–3872. doi.org/10.3892/mmr.2020.11476

19. Shinozuka, T., Terada, M., & Tanaka, E. (2006). Solid-phase extraction and analysis of 20 antidepressant drugs in human plasma by LC/MS with SSI method. *Forensic Science International*, *162*(1-3), 108–112. doi.org/10.1016/j.forsciint.2006.03.038

20. Stokes, P., & O'Connor, G. (2003). Development of a liquid chromatography–mass spectrometry method for the high-accuracy determination of creatinine in serum. *Journal of Chromatography B*, *794*(1), 125–136. doi.org/10.1016/s1570-0232(03)00424-0

21. Wang, C., Catlin, D. H., Demers, L. M., Starcevic, B., & Swerdloff, R. S. (2004). Measurement of total serum testosterone in adult men: Comparison of current laboratory methods versus liquid chromatography-tandem mass spectrometry. *The Journal of Clinical Endocrinology & Metabolism*, *89*(2), 534–543. doi.org/10.1210/jc.2003-031287

22. Li, X., Shen, B., Jiang, Z., Huang, Y., & Zhuo, X. (2013). Rapid screening of drugs of abuse in human urine by high-performance liquid chromatography coupled with high resolution and high mass accuracy hybrid linear ion trap-orbitrap mass spectrometry. *Journal of Chromatography A*, *1302*, 95–104. doi.org/10.1016/j.chroma.2013.06.028

23. Dziadosz, M., Teske, J., Henning, K., Klintschar, M., & Nordmeier, F. (2018). LC–MS/ms screening strategy for cannabinoids, opiates, amphetamines, cocaine, benzodiazepines and methadone in human serum, urine and post-mortem blood as an effective alternative to immunoassay based methods applied in forensic toxicology for preliminary examination. *Forensic Chemistry*, *7*, 33–37. doi.org/10.1016/j.forc.2017.12.007

24. Borden, S. A., Palaty, J., Termopoli, V., Famiglini, G., Cappiello, A., Gill, C. G., & Palma, P. (2020). Mass spectrometry analysis of drugs of abuse: challenges and emerging strategies. *Mass Spectrometry Reviews*, *39*(5-6), 703–744. doi.org/10.1002/mas.21624

25. Zheng, J., Mehl, J., Zhu, Y., Xin, B., &Olah, T. (2014). Application and challenges in using LC-MS assays for absolute quantitative analysis of therapeutic proteins in drug discovery. *Bioanalysis*, *6*(6), 859–879. doi.org/10.4155/bio.14.36

26. Vogeser, M., Kirchhoff, F. (2011). Progress in automation of LC-MS in laboratory medicine. *Clinical Biochemistry*, *44*(1), 4–13. doi.org/10.1016/j.clinbiochem.2010.06.005

27. Kirby, B. J., Kalhorn, T., Hebert, M., Easterling, T., &Unadkat, J. D. (2008). Sensitive and specific LC-MS assay for quantification of digoxin in human plasma and urine. *Biomedical Chromatography*, *22*(7), 712–718. doi.org/10.1002/bmc.988

28. Kailasam, S. (2021). *Liquid chromatography in the biopharmaceutical industry*. Technology Networks. Retrieved March 25, 2022, from www.technologynetworks.com/tn/articles/liquid-chromatography-in-the-biopharmaceutical-industry-354851

29. *Food safety analysis using LC/MS Equipment*. GenTech Scientific. (2021, November 30). Retrieved March 26, 2022, from https://gentechscientific.com/food-safety-analysis-using-lc-ms-equipment/

30. Rappold, B. A. (2022). Review of the use of liquid chromatography-tandem mass spectrometry in clinical laboratories: part I – development. *Annals of Laboratory Medicine*, *42*(2), 121–140. doi.org/10.3343/alm.2022.42.2.121

31. Pitt, J. J. (2009, February). *Principles and applications of liquid chromatography-mass spectrometry in Clinical Biochemistry*. The Clinical Biochemist. Reviews. Retrieved March 25, 2022, from www.ncbi.nlm.nih.gov/pmc/articles/PMC2643089/

32. Dorlo, T. P., Eggelte, T. A., de Vries, P. J., & Beijnen, J. H. (2012). Characterization and identification of suspected counterfeit miltefosine capsules. *The Analyst*, *137*(5), 1265–1274. doi.org/10.1039/c2an15641e

33. Bekoe, S. O., Bak, S. A., Björklund, E., Krogh, K. A., N. N. A. Okine, N., Adosraku, R. K., Styrishave, B., & Hansen, M. (2014). Determination of thirteen antibiotics in drug products – a new LC-MS/MS tool for screening drug product quality. *Anal. Methods*, *6*(15), 5847–5855. doi.org/10.1039/c4ay00748d

34. World Health Organization. (2019, December 4). *Malaria on the decline in who south-East Asia region; efforts must continue as risks persist: Who*. World Health Organization. Retrieved March 25, 2022, from www.who.int/southeastasia/news/detail/04-12-2019-malaria-on-the-decline-in-who-south-east-asia-region-efforts-must-continue-as-risks-persist-who

35. Centers for Disease Control and Prevention. (2022, February 2). *CDC – malaria – about malaria*. Centers for Disease Control and Prevention. Retrieved March 25, 2022, from www.cdc.gov/malaria/about/index.html

36. Wolff, J.-C., Thomson, L. A., & Eckers, C. (2003). Identification of the 'wrong' active pharmaceutical ingredient in a counterfeit Halfan™ drug product using accurate mass electrospray ionisation mass spectrometry, accurate mass tandem mass spectrometry and liquid chromatography/mass spectrometry. *Rapid Communications in Mass Spectrometry, 17*(3), 215–221. doi.org/10.1002/rcm.893

37. El-Bagary, R. I., Elkady, E. F., Mowaka, S., & Attallah, M. (2015). Validated HPLC and ultra-HPLC methods for determination of dronedarone and amiodarone application for counterfeit drug analysis. *Journal of the Association of Official Agricultural Chemists International, 98*(6), 1496–1502. doi.org/10.5740/jaoacint.15-054

38. Fejős, I., Neumajer, G., Béni, S., & Jankovics, P. (2014). Qualitative and quantitative analysis of PDE-5 inhibitors in counterfeit medicines and dietary supplements by HPLC–UV using sildenafil as a sole reference. *Journal of Pharmaceutical and Biomedical Analysis, 98*, 327–333. doi.org/10.1016/j.jpba.2014.06.010

39. Song, F., Monroe, D., El-Demerdash, A., & Palmer, C. (2014). Screening for multiple weight loss and related drugs in dietary supplement materials by flow injection tandem mass spectrometry and their confirmation by liquid chromatography tandem mass spectrometry. *Journal of Pharmaceutical and Biomedical Analysis, 88*, 136–143. doi.org/10.1016/j.jpba.2013.08.031

40. Johansson, M., Fransson, D., Rundlöf, T., Huynh, N.-H., & Arvidsson, T. (2014). A general analytical platform and strategy in search for illegal drugs. *Journal of Pharmaceutical and Biomedical Analysis, 100*, 215–229. doi.org/10.1016/j.jpba.2014.07.026

41. Hall, K. A., Pizzanelli, D., Green, M. D., Dondorp, A., De Veij, M., Mayxay, M., Fernandez, F. M., et al. (2006). Characterization of counterfeit artesunate antimalarial tablets from Southeast Asia. *The American Journal of Tropical Medicine and Hygiene, 75*(5), 804–811. doi.org/10.4269/ajtmh.2006.75.804

42. Mufusama, J.-P., Ndjoko Ioset, K., Feineis, D., Hoellein, L., Holzgrabe, U., & Bringmann, G. (2018). Quality of the antimalarial medicine artemether – lumefantrine in 8 cities of the Democratic Republic of the Congo. *Drug Testing and Analysis, 10*(10), 1599–1606. doi.org/10.1002/dta.2420

43. Centers for Disease Control and Prevention. (2021, June 23). *CDC – Malaria – diagnosis & treatment (United States) – treatment (U.S.) – artesunate.* Centers for Disease Control and Prevention. Retrieved March 25, 2022, from www.cdc.gov/malaria/diagnosis_treatment/artesunate.html

44. World Health Organization. (2022, January 8). *Leishmaniasis.* World Health Organization. Retrieved March 25, 2022, from www.who.int/news-room/fact-she ets/detail/leishmaniasis#:~:text=Visceral%20leishmaniasis%20(VL)%2C%20a lso,East%20Africa%20and%20in%20India.

45. Kessler, A., Sollie, S., Challacombe, B., Briggs, K., & Van Hemelrijck, M. (2019). The global prevalence of erectile dysfunction: a review. *British Journal of Urology International, 124*(4), 587–599. doi.org/10.1111/bju.14813

46. Erectile Dysfunction (ED): Symptoms, Diagnosis & Treatment – Urology Care Foundation. (2018). Retrieved March 26, 2022, from www.urologyhealth.org/urol ogy-a-z/e/erectile-dysfunction-(ed)

47. Sison, G. (2022, February 15). *Erectile dysfunction statistics 2020: how common is ED?* The Checkup. Retrieved March 25, 2022, from www.singlecare.com/blog/news/erectile-dysfunction-statistics/

48. Lee, J. H., Han, J. H., Kim, S., Kim, N. S., Yoon, C.-Y., Kim, J., & Baek, S. Y. (2021). Screening sexual performance enhancing compounds and their analogues in counterfeit and illicit erectile dysfunction drugs by high-performance liquid chromatography and liquid chromatography-tandem mass spectrometry. *Journal of Forensic and Legal Medicine, 82*, 102224. doi.org/10.1016/j.jflm.2021.102224

49. Fidan, A. K., & Bakirdere, S. (2016). Simultaneous determination of sildenafil and tadalafil in legal drugs, illicit/counterfeit drugs, and wastewater samples by high-performance liquid chromatography. *Journal of the Association of Official Agricultural Chemists International, 99*(4), 923–928. doi.org/10.5740/jaoacint.15-0320

50. Lee, J. H., Park, H. N., Park, O. R., Kim, N. S., Park, S.-K., & Kang, H. (2019). Screening of illegal sexual enhancement supplements and counterfeit drugs sold in the online and offline markets between 2014 and 2017. *Forensic Science International, 298*, 10–19. doi.org/10.1016/j.forsciint.2019.02.014

51. Song, F., El-Demerdash, A., & Lee, S.-J. S. (2012). Screening for multiple phosphodiesterase type 5 inhibitor drugs in dietary supplement materials by flow injection mass spectrometry and their quantification by liquid chromatography tandem mass spectrometry. *Journal of Pharmaceutical and Biomedical Analysis, 70*, 40–46. doi.org/10.1016/j.jpba.2012.05.017

52. Haneef, J., Shaharyar, M., Husain, A., Rashid, M., Mishra, R., Siddique, N. A., & Pal, M. (2013). Analytical methods for the detection of undeclared synthetic drugs in traditional herbal medicines as adulterants. *Drug Testing and Analysis, 5*(8), 607–613. doi.org/10.1002/dta.1482

53. Venhuis, B. J., Zomer, G., Vredenbregt, M. J., & De Kaste, D. (2010). The identification of (−)-trans-tadalafil, tadalafil, and sildenafil in counterfeit Cialis and the optical purity of tadalafil stereoisomers. *Journal of Pharmaceutical and Biomedical Analysis, 51*(3), 723–727. doi.org/10.1016/j.jpba.2009.08.010

54. Venhuis, B. J., Zwaagstra, M. E., Keizers, P. H. J., & De Kaste, D. (2014). Dose-to-dose variations with single packages of counterfeit medicines and adulterated dietary supplements as a potential source of false negatives and inaccurate health risk assessments. *Journal of Pharmaceutical and Biomedical Analysis, 89*, 158–165. doi.org/10.1016/j.jpba.2013.10.038

55. Sizar, O., & Pico, J. (2021, May 21). *Androgen replacement*. StatPearls [Internet]. Retrieved March 25, 2022, from www.ncbi.nlm.nih.gov/books/NBK534853/

56. Myers, J. B., & Meacham, R. B. (2003). *Androgen replacement therapy in the aging male*. Reviews in urology. Retrieved March 25, 2022, from www.ncbi.nlm.nih.gov/pmc/articles/PMC1508369/

57. Handelsman, D. J. (2021, July 16). *Androgen misuse and abuse*. Endocrine reviews. Retrieved March 25, 2022, from https://pubmed.ncbi.nlm.nih.gov/33484556/

58. Cho, S.-H., Park, H. J., Lee, J. H., Do, J.-A., Heo, S., Jo, J. H., & Cho, S. (2015). Determination of anabolic–androgenic steroid adulterants in counterfeit drugs by UHPLC–MS/MS. *Journal of Pharmaceutical and Biomedical Analysis, 111*, 138–146. doi.org/10.1016/j.jpba.2015.03.018

59. Fiori, J., & Andrisano, V. (2014). LC–MS method for the simultaneous determination of six glucocorticoids in pharmaceutical formulations and counterfeit cosmetic products. *Journal of Pharmaceutical and Biomedical Analysis, 91*, 185–192. doi.org/10.1016/j.jpba.2013.12.026

60. Savaliya, A. A., Prasad, B., Raijada, D. K., & Singh, S. (2009). Detection and characterization of synthetic steroidal and non-steroidal anti-inflammatory drugs in Indian Ayurvedic/herbal products using LC-MS/TOF. *Drug Testing and Analysis*, *1*(8), 372–381. doi.org/10.1002/dta.75

61. World Health Organization. (2021, June 9). *Obesity and overweight*. World Health Organization. Retrieved March 25, 2022, from www.who.int/news-room/fact-she ets/detail/obesity-and-overweight

62. Centers for Disease Control and Prevention. (2021, September 30). *Adult obesity facts*. Centers for Disease Control and Prevention. Retrieved March 25, 2022, from www.cdc.gov/obesity/data/adult.html

63. Lee, J. H., Park, H. N., Kim, N. S., Park, H.-J., Park, S., Shin, D., & Kang, H. (2019). Detection of illegal abortion-induced drugs using rapid and simultaneous method for the determination of abortion-induced compounds by LC–MS/MS. *Chromatographia*, *82*(9), 1365–1371. doi.org/10.1007/s10337-019-03758-1

64. Abdel-Megied, A. M., & Badr El-din, K. M. (2019). Development of a novel LC–MS/MS method for detection and quantification of tramadol hydrochloride in presence of some mislabeled drugs: application to counterfeit study. *Biomedical Chromatography*, *33*(6). doi.org/10.1002/bmc.4486

65. Pang, W., Yang, H., Wu, Z., Huang, M., & Hu, J. (2009). LC-MS–MS in MRM mode for detection and structural identification of synthetic hypoglycemic drugs added illegally to 'natural' anti-diabetic herbal products. *Chromatographia*, *70*(9-10), 1353–1359. doi.org/10.1365/s10337-009-1344-0

66. World Health Organization. (2017, September 28). *Women and girls continue to be at risk of unsafe abortion*. World Health Organization. Retrieved March 25, 2022, from www.who.int/reproductivehealth/topics/unsafe_abortion/abortion-safety-estimates/en/

67. Vanhee, C., Janvier, S., Desmedt, B., Moens, G., Deconinck, E., De Beer, J. O., & Courselle, P. (2015). Analysis of illegal peptide biopharmaceuticals frequently encountered by controlling agencies. *Talanta*, *142*, 1–10. doi.org/10.1016/j.talanta.2015.04.022

68. Han, S.-Y., Zhou, J., Lian, H.-Z., & Tan, L. (2013). Identification of potentially counterfeit ingredient in 4-chlorodehydromethyl testosterone tablets by LC-ESI-MS. *Asian Journal of Chemistry*, *25*(1), 147–151. doi.org/10.14233/ajchem.2013.12841

4 Analyzing Counterfeit and Falsified Herbal Products with GC/MS

Maryam Akhgari[1]and Afshar Etemadi-Aleagha[2]*
[1]Legal Medicine Research Center, Legal medicine Organization, Tehran, Iran
[2]Department of Anesthesiology and Critical Care, Amir-Alam Hospital, Tehran University of Medical Sciences, Tehran, Iran

CONTENTS

4.1 Introduction 98
4.2 Adulteration of herbal drugs...99
    4.2.1 Reasons for Adulteration 99
    4.2.2 Various Methods of Adulteration of Herbal Drugs 99
4.3 Types of Analysis for Authentication Identification.............................99
    4.3.1 Detection of Active Pharmaceutical Ingredients in Counterfeit Herbal Drugs 100
    4.3.2 Sample Preparation Prior to Analytical Toxicology Process.........100
        4.3.2.1 Microextraction Techniques...101
        4.3.2.2 Liquid Phase Extraction (LPE) 101
        4.3.2.3 Pressurized Liquid Extraction (PLE) 101
        4.3.2.4 Dispersive Liquid-Liquid Microextraction (DLLME) 101
        4.3.2.5 Ultrasonic Assisted Extraction..102
        4.3.2.6 Microwave Assisted Extraction (MAE).........................102
        4.3.2.7 Dispersive Liquid-Liquid Microextraction (DLLME) 102
        4.3.2.8 Hollow Fiber-Based Liquid-Phase Microextraction 102
        4.3.2.9 Solid Phase Extraction (SPE)..102
        4.3.2.10 Solid Phase Microextraction (SPME)...........................103
        4.3.2.11 Dispersive Solid-Phase Extraction (d-SPE)..................103
        4.3.2.12 QuEChERS 103
    4.3.3 Instrumental Methods for Undeclared Drug Detection in Counterfeit Herbal Drugs...103
        4.3.3.1 Spectrophotometric Techniques.....................................104
        4.3.3.2 Chromatographic Techniques 104

DOI: 10.1201/9781003270461-4

4.4   Gas Chromatography/Mass Spectrometry for Counterfeit
        Herbal Drugs Detection..109
4.5   Comparing GC/MS and HPLC in the Analysis of Counterfeit
        Herbal Drugs 110
4.6   Necessity for Method Validation in Counterfeit Drug Analysis..............110
        4.6.1   Method Validation 111
                4.6.1.1   Selectivity...111
                4.6.1.2   Accuracy 111
                4.6.1.3   Precision...111
                4.6.1.4   Limit of Detection (LOD).....................................111
                4.6.1.5   Limit of Quantification (LOQ)..............................111
                4.6.1.6   Linearity...111
                4.6.1.7   Recovery of Extraction Efficiency 111
4.7   Conclusion...112
References...112

4.1 INTRODUCTION

Herbal drug preparations have been used for centuries. Due to their beneficial, economical, and widespread use in developed and developing countries, herbal drugs are sometimes found to be adulterated and falsified (1). Many physicians have warned about the dangers of counterfeit drugs (2). Most of the drug falsifications that have been reported in herbal preparations resulted in unwanted side effects due to toxicity or lack in capacity to manage or treat the diseases (3). Drug counterfeiting is a growing problem, especially as a result of the availability of social media and internet sources. The rise in clandestine trade in counterfeit and falsified herbal products is a major threat to public health. The World Health Organization (WHO) defines falsified medicines as drugs "that deliberately/ fraudulently misrepresent their identity, composition or source" (4). Falsified drugs with incorrect and inactive ingredients or with inaccurate amounts of the active agent is one of the big challenges for health authorities in many countries (5). It is among the main claims referred to forensic practitioners due to adverse drug events with severe clinical manifestations in counterfeit drug users. Counterfeit drugs are prepared prior to chemical analysis using different techniques in order to optimize instrumental analysis results. Forensic toxicology and chemistry are fields in which several analytical techniques are used for the analysis of various types of samples for the detection of drugs and poisons.

We begin this chapter with a discussion of the basic concepts and definition of counterfeit herbal drugs, followed by discussion of different sample preparation and instrumental analysis methods for their characterization in forensic toxicology laboratories, and we end the chapter with a discussion of the role of gas chromatography/mass spectrometry (GC/MS) as an efficient tool for counterfeit drugs analysis, its basic concept and definition.

4.2   ADULTERATION OF HERBAL DRUGS

4.2.1   REASONS FOR ADULTERATION

Any kind of drugs that are in high demand by patients and have a highly profitable market are gaining popularity for targeted adulteration and counterfeiting (6, 7). As a result of low production cost, increasing demand, and easy access through online sources, the counterfeit drug trade is becoming a lucrative business all over the world (6). Due to the different regulation and drug policy in disparate countries, counterfeit drugs are produced and marketed globally with high turnover profit (8).

There is an idea among patients that herbal drugs have little or no side effects in comparison to synthetic drugs. As achieving to the desired goal cannot be obtained rapidly, many herbal drug producers adulterate herbal preparations with synthetic drugs to get results in a short period of time. As a result, there is a tremendous surge in the sale and marketing of these kinds of drugs. The other reason for herbal drug adulteration is economic motivation (9).

Typically adulterated herbal products with an extensive global market that are deemed "natural drugs" include those for losing weight (6,10-12), addiction cessation (13), erectile dysfunction treatment (6,14,15), as well as increasing body mass, mainly for use by body builders (16).

4.2.2   VARIOUS METHODS OF ADULTERATION OF HERBAL DRUGS

Adulteration is performed by adding other materials (such as wrong and cheaper plant materials or, conversely, the more expensive synthetic drugs) to herbal medicines to boost profit. Adding synthetic drugs to herbal medicine is becoming a common practice in some countries. Adulterated herbal drugs used as a substitute for prescription drugs represent a growing issue for global public health (13).

This chapter explores the evidence related to adulterated herbal drugs including their forensic aspects, profiling, and laboratory analysis.

Several analytical methods are available for chemical and pharmaceutical analysis of suspected counterfeit products. There are typically three steps in an analysis procedure for the detection of active ingredients in counterfeit medicines: (1) sample preparation, (2) sample analysis, and (3) interpretation of obtained results.

4.3   TYPES OF ANALYSIS FOR AUTHENTICATION IDENTIFICATION

The WHO's definition of falsified and counterfeit drugs highlights many factors in drug packaging. There are some distinctive types of drug analysis for the discrimination between genuine and falsified drugs; namely verification of the supplier (license from health authorities), visual package inspection (labeling, package insert), active pharmaceutical ingredients (APIs) ingredient name, dosage form, manufacturer detail, batch and lot number, production and expiry dates, holograms, physical (content uniformity, weight, hardness, friability), and, finally, chemical analysis (TLC, HPLC, GC/MS, LC/MS, and other spectrophotometric methods) (17).

4.3.1 DETECTION OF ACTIVE PHARMACEUTICAL INGREDIENTS IN COUNTERFEIT HERBAL DRUGS

Discriminating between authentic and counterfeit drugs is not always a trivial task. Differentiation between these two product categories needs to compare features on packaging, labeling, leaflet inserts, and physical examination of the dosage form (18). These drugs are often designed to appear identical to the original drugs, therefore, visual inspection is inadequate and must be confirmed by laboratory analysis (19). Many drug manufacturers use marketing, coding systems, or trademarks to check counterfeiting in order to protect their brand, product, and reputation as well as to facilitate fast, near effortless recognition by law enforcement personnel (6). However, it is logical to obtain information on chemical composition and APIs in counterfeit products in order to access the potential health hazard associated with counterfeit products. Herbal preparations are frequently advertised as dietary supplements and do not go through robust analysis although they may contain undeclared APIs. Multiple drug classes, including legal and illegal drugs, are deceptively added to herbal preparations. Many pharmaceutical dosage forms are also prone to falsification. Capsules, powders (in sachets), and liquid dosage forms (syrups, juices, etc.) are regularly falsified. As a result of the broad range of APIs in herbal preparations, sensitive and specific analytical techniques must be used in order to qualitatively and quantitatively determine the chemical constituents of fake drugs.

4.3.2 SAMPLE PREPARATION PRIOR TO ANALYTICAL TOXICOLOGY PROCESS

Most biological and non-biological matrices, subjected to trace drug analysis carry a high matrix effect that can interfere with instrumental detection. Choosing a suitable sample preparation procedure is dependent on many factors, including volatility of the analyte, matrix type, and chemical structure of the drug (20). With the demand for decreasing detection limits by legal authorities and public regulations, sample preparation and extraction of analytes from different matrices are among the priorities of forensic toxicology laboratories. Currently there is a trend toward the use of alternative and automated forms of sample preparation and concentration methods, such as thermo-desorption, headspace and purge, and trap techniques that are coupled to high-tech instruments (21).

Optimal drug extraction from complex matrices can reduce sources of error and analysis time. Efficient sample preparation enables unequivocal qualitative and quantitative analysis (21). Sample preparation is the first step in an analytical toxicology procedure. Sample clean-up as well as analyte extraction and enrichment is crucial in toxicological analysis, especially in biological matrices and pharmaceutical formulations, due to matrix interferences and trace amounts of analytes (21).

Different extraction techniques are employed in order to reduce matrix effects in biological and non-biological matrices. The presence of compounds such as salts, acids, proteins, lipids, and other organic compounds may interfere in drug analysis. Some drugs of forensic interest that are considered as target drugs include, but are not limited to, licit and illicit drugs, poisons, and alcohol (22). Common sample preparation methods in forensic toxicology practice include liquid-liquid extraction (LLE), solid phase extraction (SPE), and novel preparation procedures based on microextraction that have been developed and used in forensic toxicology practice, using a minimum amount of solvents compared to conventional methods. Microextraction techniques are more cost effective with higher extraction yield and recoveries. Moreover, they are environmentally safe and user friendly (22).

Apart from sample preparation, the development and validation of an analytical method is of great importance to obtain desired results.

4.3.2.1   Microextraction Techniques

Microextraction methods are defined as sample preparation procedures in which the volume of extracting solvent is very small in comparison to the biological and non-biological matrices (22). Modern extraction techniques regarded as green methods, such as pressurized liquid extraction (PLE) and microwave assisted extraction (MAE), have been subject to increased attention in analytical chemistry due to their high extraction yields, selectivity, reduction in use of organic solvents, and safety aspects (23).

4.3.2.2   Liquid Phase Extraction (LPE)

Liquid phase extraction is one of the oldest methods for the isolation of organic compounds using organic solvents to extract target drugs or analytes from solid matrices by automated or manual shaking. Disadvantages of LPE includes being time consuming and typically requiring large volumes of organic solvents (24).

4.3.2.3   Pressurized Liquid Extraction (PLE)

Pressurized liquid extraction is performed in high temperature (50–200°C) and pressure (1000–3000 psi) to improve extraction efficiency. PLE is more efficient for the preparation of solid and semisolid samples in comparison to LPE. Moreover lower amounts of extracting solvents would be used (25, 26). If there is needed to have good recovery and higher extraction efficiency, PLE can be coupled with solid phase extraction (SPE) (24).

4.3.2.4   Dispersive Liquid-Liquid Microextraction (DLLME)

DLLME is a tertiary component extraction and sample preparation technique that involves the dispersion of extraction solvent droplets in an aqueous medium.

It provides a large surface contact between the sample and dispersed droplets. Extraction equilibrium can be obtained between the small droplets and aqueous sample; therefore, enrichment of the analyte can be achieved easily in extractant phase. Droplets can be subsequently separated from the aqueous medium after centrifugation. This technique has high recovery and extraction efficiency for drugs and organic matters in liquid phases. DLLME employs smaller amounts of organic solvents and has a shorter preparation time in comparison to SPE. Also, it shows higher extraction recovery and better enrichment factors (27).

4.3.2.5   Ultrasonic Assisted Extraction

In ultrasonic assisted extraction, the process of inserting target drugs into the extracting solvent in LPE is accelerated by applying ultrasound. Many physicochemical phenomena – such as ultrasonic disturbance, cavitation, erosion, emulsification, and fragmentation – improve extraction efficiency and quality compared to normal extraction procedures using organic solvents (28).

4.3.2.6   Microwave Assisted Extraction (MAE)

Microwave assisted extraction is a process in which microwave energy is used to heat solvents in contact with a matrix containing drugs in order to promote partitioning of the compounds from the sample into the extracting solvent in a short period of time and as a result of heating up the solution (29).

4.3.2.7   Dispersive Liquid-Liquid Microextraction (DLLME)

In conventional LLE, large amounts of organic solvents are used that may not be environmentally safe. LLME methods were developed to reduce organic solvents consumption as well as boost the extraction efficiency (30). DLLME and hollow fiber-based liquid-phase microextraction are two routine LLME techniques that are used for the preparation of liquid samples prior to analysis.

4.3.2.8   Hollow Fiber-Based Liquid-Phase Microextraction

A typical hollow fiber-based liquid-phase microextraction consists of a porous hollow fiber membrane which is impregnated with an organic supported liquid membrane. The extraction solvent is injected into the lumen of the fiber. The fiber is immersed in the liquid sample for the extraction of the desired analytes. This extraction method has some advantages compared with other liquid phase microextraction, including being stable, low cost, superior clean up effects, and good extraction yields (31).

4.3.2.9   Solid Phase Extraction (SPE)

Solid phase extraction (SPE) is used as an efficient sample preparation technique in food chemistry, pharmaceutical, and forensic drug analysis and follows a four-step procedure: (1) column activation or conditioning, (2) sample loading, (3) clean-up impurities by washing the adsorbent, and (4) analyte elution (32). This allows for trace amounts of target analytes to be concentrated in a solid phase.

Moreover, SPE can reduce the sample matrix effect efficiently (33). SPE has many advantages compared with traditional liquid extraction methods, including less solvent consumption and shorter extraction time. SPE is considered as one of the best sample preparation techniques for liquid samples due to the large number of developed solid sorbent phases and automated steps (34).

4.3.2.10   Solid Phase Microextraction (SPME)

Solid phase microextraction (SPME) involves the use of a syringe whose needle contains a silica fiber coated with an extracting phase that can be either a liquid (polymer) or solid (sorbent). The needle can be immersed directly in a liquid sample or placed in the top for headspace extraction from complex matrices with high matrix effects (33).

4.3.2.11   Dispersive Solid-Phase Extraction (d-SPE)

Solid phase extraction is one of the most widely used sample preparation methods. However, there are some disadvantages such as stacking packing sorbent material in SPE cartridges. To combat this problem, dispersive solid phase extraction (d-SPE) was developed. Instead of packing the sorbent in cartridges, a small amount of sorbent is added to the liquid containing analyte and then separated from the sample. Low consumption of sorbent, user friendliness, and prevention of sorbent blockage are among the advantages of d-SPE (35).

4.3.2.12   QuEChERS

QuEChERS is featured as a "quick, easy, cheap, effective, rugged, and safe" extraction method that has been developed for trace extraction of analytes from complex matrices, such as food and biological samples. QuEChERS is valued for its low cost, simplicity, high extraction yield, and ability to extract drugs from various matrices. It also affords flexibility in the type of instrumental analytical technique used (36). In this four-step sample preparation method, extraction of an active ingredient is achieved by the use of trace amounts of organic solvents, inorganic salts – such as magnesium sulfate ($MgSO_4$) – and a dispersion sorbent for sample clean-up (36). QuEChERS-dSPE was developed for the rapid screening of illegal adulterants in soft-gel-type supplements for the detection of undeclared active pharmaceutical ingredients in erectile dysfunction and weight loss drugs using ultra-high performance liquid chromatography quadrupole/time of flight-mass spectrometry (UHPLC-Q/TOF-MS) (37).

4.3.3   INSTRUMENTAL METHODS FOR UNDECLARED DRUG DETECTION IN COUNTERFEIT HERBAL DRUGS

Detailed knowledge of chemical constituents of herbal based preparations (tablets, capsules, powders, liquids, herbal distillates, etc.) has been facilitated using powerful instrumental techniques. This section summarizes the application of sample preparation and instrumental techniques for drug characterization in toxicology laboratories.

4.3.3.1   Spectrophotometric Techniques

Spectrophotometric techniques are preferably used for the identification of counterfeit drugs due to the fast and specific detection of substances with minor sample preparation procedures. Spectrophotometric techniques include infrared (IR), fluorescence, UV-visible, Raman spectroscopy, nuclear magnetic resonance (NMR) (38), Fourier-transform infrared spectroscopy (FTIR), near-infrared spectroscopy (NIR) (38). Raman spectroscopy and FTIR are among commonly used methods for the determination of counterfeit drugs (38-40). ATR-FTIR spectroscopy and chemometrics are two analytical methods used for the determination of residual solvents like ethanol, ethyl acetate, acetone, and chloroform in counterfeit drugs in comparison to original tablets (38,41). Each technique has its own specification in the detection of drugs in forensic toxicology practice.

4.3.3.2   Chromatographic Techniques

Chromatographic instrumentations are widely used in forensic toxicology laboratories for the characterization of adulterated drugs. In the majority of published works, chromatography is used for the isolation and separation of APIs from dosage forms, or for detection thereof in the medicines. For example, chromatographic techniques are used to detect APIs in traditional Chinese herbal medicines (38). Chromatography can simply be described as a separation procedure that consists of mobile and stationary phases. The mobile phase is the part that contains analyte, while the stationary phase is the fixed phase that separate drugs using different mechanisms. Depending on the design of the chromatography system, both phases can be solid, liquid, or gas.

A number of advanced technology instruments based on chromatography have launched and been developed for the detection of hidden APIs in herbal preparations. A wide variety of analytical techniques are applied for the qualitative and quantitative chemical analysis, as well as the profiling of counterfeit and falsified drugs. Some robust, effective, and precise examples of these methods include thin layer chromatography (TLC), Fourier-transform infrared spectroscopy (FTIR), infrared spectroscopy (IR spectroscopy), capillary electrophoresis (CE), high performance liquid chromatography (HPLC), gas chromatography-mass spectrometry (GC/MS), liquid chromatography with tandem mass spectrometry (LC/MS/MS), near-infrared spectroscopy (NIRS), Raman spectroscopy, and Nuclear magnetic resonance spectroscopy (NMR spectroscopy) (5, 8, 24, 42). The proper choice of method allows for fast, accurate, and reliable answers to questions regarding verification of the genuineness of the drug product, detecting hidden APIs, and geographic location of drug producers. As many of these instrumental methods have specific advantages, they should be used in such a way to make the most of their potential and mitigate their constraints.

4.3.3.2.1 Thin Layer Chromatography (TLC)

TLC is widely used for qualitative and quantitative analysis of drugs in different dosage forms. One of the main drawbacks of this method is the use of large amounts of organic solvents. Also, manual TLC methods suffer from variation in reproducibility and accuracy as a result of operator experience in performing the procedure. Notably, high performance thin layer chromatography (HPTLC) has higher reproducibility and accuracy; however, this technique cannot be widely used in many laboratories due to the high cost of chromatographic plates (5).

4.3.3.2.2 Liquid Chromatography Technique (LC)

Liquid chromatography (LC) is one of the most commonly used analytical techniques that can separate and detect APIs in different mixtures, both qualitatively and/or quantitatively. Separation of the analytes takes place when the solvents (mobile phase) move through the stationary phase. Liquid chromatography can be applied for the analysis of different drug classes, pesticides, toxins, and so forth. (43). The stationary phase is made up of small particles that are densely packed in a column, while the mobile phase is pushed through the stationary phase for the separation of analytes. Different kinds of samples can be analysed using HPLC with high precision, great sensitivity, and resolution. Some drawbacks associated with HPLC technique are high cost, extensive sample preparation and processing steps, time consuming, and the requirement of trained operators (44). By using new technical advances in different components of the instrument, ultra-HPLC was designed and has gained attention in forensic toxicology practice (45).

HPLC remains one of the gold standard techniques for the detection of APIs in counterfeit drugs. Saberi *et al.*, detected cyproheptadine, dexamethasone, sildenafil, tramadol, caffeine, and acetaminophen as hidden APIs in herbal weight gain supplements using ultra-HPLC and GC/MS instrumentations (16). Adulterants were shown to be deliberately added to herbal sexual activity enhancer drugs when analyzed by Hafizi *et al.*, (14). HPLC, which is based on reversed phase chromatography, was also used by Rozet *et al.*, for the detection of five analogues of phosphodiesterase inhibitors in fake Viagra tablets (15).

4.3.3.2.3 Gas Chromatography/Mass Spectrometry (GC/MS)

GC/MS is an analytical technique that combines the features of gas chromatography and mass spectrometry for the qualitative and quantitative detection of drugs and substances and dates back to the 1960s. Gas chromatography is an important method for the chemical analysis of low molecular weight substances. Additionally, as a result of the development of column technology for the production of temperature-stable phases, it is possible to detect stable compounds and drug molecules in elution temperatures of up to 500°C (20). Fused silica GC columns were introduced and reliable GC/MS instruments were manufactured in the 1980s. In 1990, automated high speed GC/MS with high separation power was introduced to the science of analytical chemistry (46). This combination is compatible with almost all sample preparation techniques (21).

Mass spectrometry is used as a detection method that determines substance molecules or its fragments. Obtained results are used as confirmation tool for drug detection. GC/MS is now routinely used for rapid analysis with high selectivity and sensitivity for a broad spectrum of applications, such as drug analysis in forensic practice (20).

4.3.3.2.3.1 Gas Chromatography/Mass Spectrometry Principles

The main principle of gas chromatography is the volatilization of the prepared sample in a heated chamber prior to the separation of the analyte or analytes in the column and, later, detection using a detector. GC as a separation technique has high efficiency and good properties for combination with mass spectrometry. Volatile or semi-volatile analytes that are thermally stable can be separated and analysed using GC. Polar molecules tend to need higher temperatures to volatilize due to strong intramolecular forces. This high temperature may inadvertently cause drug decomposition on the GC column (21). Highly polar small molecules–such as salts of quaternary ammonium– cannot be analysed directly using GC. Chemical derivatization, as a method for increasing sensitivity and selectivity, converts polar molecules to less polar substances that can be analyzed using GC due to the change in vapor pressure and absorption characteristics. As with other chromatography instrumentations, GC suffers from long analysis time. Fast GC uses improvement in optimized run conditions to provide faster analysis times with acceptable resolution. A number of analysis parameters can be manipulated in order to perform faster GC analysis. Stationary phase composition, film thickness, column length and internal diameter, carrier gas flow, and oven temperature can be individually changed and modified (21). When volatile organic compounds are separated in a GC column, detection is equally important, if not more so (46). Coupling of GC with a proper detector makes it a powerful instrument in forensic chemistry.

4.3.3.2.3.2 GC Detectors

Various detectors for coupling with GC have been developed. Some common GC detectors are: flame ionization detector (FID), nitrogen phosphorous detector (NPD), electron capture detector (ECD), flame photometric detector, and mass spectrometer (MS). The choice of detector is dependent on the field of analytical chemistry and toxicology it is being employed in and whether it is to provide qualitative or quantitative data. (46). GC coupled with MS (i.e., GCMS) is an analytical instrument that has been widely used for counterfeit drugs characterization (8). Counterfeit Ritalin tablets from California were analyzed using both GC/FID and GC/MS. Oxycodone and dihydrocodeinone were found to be adulterants in these methylphenidate tablets (47). Soltaninejad et al., used GC/MS and HPLC as methods of choice to characterize counterfeit Temgesic vials. Results of their study confirmed that fake Temgesic vials were adulterated with heroin, acetylcodeine, and pheniramine (48). Counterfeit Viagra tablets in Hungary were analyzed utilizing GC/MS, revealing that they contained amphetamine rather than sildenafil (49). Both GC/MS and LC/MS were used for the qualitative and quantitative analysis of seized drugs from the informal market among body builders in France. Anabolic androgenic steroids were undeclared APIs in these performance and image enhancing drugs (50).

Depending on carrier gas, oven temperature and column specification, GC/MS provides highly efficient and specific separation and detection of drug analytes. Capillary columns are made from thin fused silica capillaries with 10–100 m length and a 250 μm inner diameter. The stationary phase is coated on the inner surface and is selected based on the polarity of the analyte. A crucial rule that must be taken into account in choosing stationary phases is that the polarity of the stationary phase should be similar to that of the solute. Open tubular columns consist of cyanopropylphenyl dimethyl polysiloxane, carbowax polyethyleneglycol, biscyanopropyl cyanopropylphenyl polysiloxane, and diphenyl dimethyl polysiloxane (21).

Inert gasses, such as nitrogen or helium, are used to carry sample containing drug molecules through the specialized column coated with the stationary phase. The separation rate of analyte in the column is dependent on the affinity of the analyte toward the stationary and mobile phases. The higher the affinity of a drug molecule to the mobile phase, the faster the molecule moves through the column (8). Sample injection is performed through manual or autosampler to the column (44). Finally, different types of stationary phases are used for the detection of hidden APIs in counterfeit herbal drugs.

4.3.3.2.2.1   Mass Spectrometer Instrumentation

A mass spectrometer detector consists of five distinct parts: (1) an ion source, (2) a vacuum system, (3) mass analyzer, (4) detector, and (5) a computer.

4.3.3.2.2.1.1   Ion Source

After the preparation procedure is completed for testing counterfeit drugs, the eluted and isolated gaseous APIs are introduced to an ion source by continuous flow of carrier gas. It is then ionized and accelerated into the mass analyzer.

All mass techniques require an initial ionization step (51). Ionization can be accomplished by the addition of one or more protons to the molecule. Positively charged ions are formed as a results of protonation process. Produced ions have a greater mass than the neutral molecule by the number of protons that have been added. In the negative ion mode of a mass operation, a proton is lost from the drug molecule, or when negatively charged moieties, such as those in the hydroxyl group, are added to the drug structure. These two processes are referred to as chemical ionization (CI). Another process of ion production is the removal of one or more electrons from the drug molecule using electron ionization (EI). This method of ionization is the most commonly performed approach in GC/MS instruments (20). Some other ionization methods are inductivity coupled plasma (ICP), MALDI, and atmospheric pressure matrix-assisted laser desorption ionization. Produced ions travel from their ion source to the mass analyzer with the aid of an ion trap system, such as a vacuum pump, to remove interfering materials – such as residual particles and non-ionized compounds – from the ion beam (20).

Separation of ions is performed according to their m/z values in the mass analyzer. Additionally, when energy is imported into the ionized molecules, they may undergo fragmentation, producing independent and unconnected chemical species. The unfragmented form of the intact molecule is defined as its molecular ion, or "mother ion". Molecular ion fragmentation products are referred to as fragment ions, or "daughter ions". If soft fragmentation of a molecule occurs and little fragmentation takes place, the most abundant peak in the mass spectrum is often the molecular ion and is called base peak, with the relative abundance value being placed at 100%. Relative abundance of each ion is plotted as a function of m/z in mass spectrum. However, if extensive and hard fragmentation of the molecules occurs, one of the fragment ions may be more abundant and thus denoted as the base peak in the resulting spectra. Obtained intensities and m/z ratios can be used for quantitative and qualitative determination of APIs in counterfeits and/or herbal drugs.

4.3.3.2.2.1.1.1 *Electron Ionization (EI)*   Ion production can be accomplished via the bombardment of gas-phase molecules by electrons with a kinetic energy of 70 eV emitted from a heated filament and attracted to a collector electrode (51). A vacuum pump prevents filament oxidation and minimizes scattering of electron beam (51).

EI is referred to as a hard ionization method due to the propensity for the production of fragment ions. Patterns are unique for each substance and used as a fingerprint for the identification of drugs by matching obtained mass spectra of unknown analytes to entries in mass spectral libraries. The fragmentation process, ions distribution, and the extent of molecular ions production are dependent on the compound's chemical structure, bonds, and stability (20). This approach allows for relatively rapid quantitative analysis.

4.3.3.2.2.1.1.2 *Chemical Ionization (CI)*   In comparison, CI is a softer ionization technique. Gas-phase molecules are ionized by transferring or abstracting a proton mediated by gas molecules, such as isobutene, water, ammonia, or methane. In this ionization method, little fragmentation occurs and a protonated version of the natural molecules are produced as molecular ions (51). Since the molecular ion + 1 peak is most often the base peak, quantification of the initial compound is possible with this approach.

4.3.3.2.2.1.2 Mass Analyzer
In mass analyzers, electromagnetic fields are used for the separation and measurement of ions according to their m/z. In GC/MS systems, single mass analyzers – such as magnetic sector, quadrupole, ion-trap, and time-of-flight

(TOF) – are used. Quadrupole mass analyzers are employed extensively in GC/MS instruments. This type of mass analyzer consists of four parallel cylindrical metal rods. The quadrupole mass analyzer scans, transmits, and sequentially detects ions from the lowest to the highest m/z in the selected m/z range (51).

4.3.3.2.2.1.3 Detectors

Detectors are used for the identification and quantification of analytes in a mixture. Most GC/MS instruments use electron multipliers for the detection of selected ions. Discrete and continuous dynode electron multipliers are two main types in mass spectrometry. Both multipliers amplify the response and generate more electrons and higher electric current. Amplified signals are measured for each m/z and recorded in mass spectrum. An electron multiplier also reduces excessive noise and redundant ions to obtain higher sensitivity (20).

4.3.3.2.2.1.4 Computer

Library searches for the identification of drugs is an important function in toxicology laboratories. Several commercial and self-generated libraries are used for spectral matching. Many government agencies make libraries freely accessible to assist those that are working to mitigate the counterfeit drug crisis.

4.3.3.2.4 *Liquid Chromatography/Tandem Mass Spectrometry (LC/MS/MS)*

LC/MS/MS is another common instrument for the detection of drug counterfeiting. The method is fast and sensitive enough for the analysis of large number of drugs in a short period of time. In a study conducted by Label *et al.*, 71 counterfeit erectile dysfunction drugs and 11 natural excipients were analyzed using mass spectrometer in ~10 minutes (52). Additionally, six corticosteroids, identified as banned drugs, were analyzed in cosmetic and natural products as adulterant using UHPLC/MS/MS (53). Likewise, ultra-performance liquid chromatography (UPLC) with diode array detectors, coupled with electroscopy ionization quadrupole time of flight (Q-TOF) mass spectrometry, was used for the analysis of 43 erectile dysfunction medicines and 65 fake drugs. In most of the analyzed samples, APIs other than those indicated on the packages were detected. Also, the concentration of detected APIs was higher compared to typical drugs (54). Cho *et al.*, developed and validated the UHPLC/MS/MS method for the detection of 26 anabolic steroids quantitatively in counterfeit drugs used as dietary supplements and drugs for the improvement of muscle mass (55).

4.4 GAS CHROMATOGRAPHY/MASS SPECTROMETRY FOR COUNTERFEIT HERBAL DRUGS DETECTION

GC/MS has become one of the most commonly utilized instruments for the characterization of counterfeit drugs. For example, seized counterfeit sildenafil tablets from Hungary were analyzed and shown to contain amphetamine, which was detected using GC/MS (56). Likewise, oxycodone and dihydrocodone were two undeclared APIs in counterfeit Ritalin tablets (47). Liu *et al.*, (57) proposed GC/MS and liquid chromatography-diode array detector (LC-DAD) methods for the detection of active ingredients as adulterants in herbal products advertised as natural slimming drugs. The most common drug categories were anorexics, antidepressants, anxiolytics, and diuretics. In a study conducted on 42 anabolic steroids obtained from an illegal German market, GC/MS analysis showed that 15 did not contain any active ingredients (58). Hafizi *et al.,* analyzed different dosage forms of herbal sexual activity enhancer drugs using GC/MS and HPLC and found sildenafil, tramadol, and diazepam as hidden APIs in the herbal preparations (14). Saberi *et al.*, analysed herbal weight gain supplements using GC/MS and HPLC quantitatively. They reported that the amount of dexamethasone and cyproheptadine were higher than therapeutic doses (16). In Jordan, GC/MS was applied for drug detection in Captagon (fenethylline) tablets. The counterfeit tablets contained amphetamine, methamphetamine, antimalarial drugs, antibiotics, and sympathomimetics (59). Herbal weight loss drugs found in Iranian markets were analyzed using GC/MS. Tramadol, caffeine, fluoxetine, rizatriptan, venlafaxine, and methadone were detected hidden APIs (60). Neves *et al.*, performed GC/MS for the detection of anabolic steroids in counterfeit medicines and dietary supplements in Brazil using GC/MS (61). This analytical technology was used as a confirmatory method for quantitative analysis of counterfeit herbal medicines used as opioid substitution therapy. Results showed that these herbal supplements contained diphenoxylate, tramadol, acetaminophen, codeine, sertraline, and fluoxetine (13).

4.5   COMPARING GC/MS AND HPLC IN THE ANALYSIS OF COUNTERFEIT HERBAL DRUGS

GC/MS and HPLC are different chromatographic methods that are widely used in the quantitative and qualitative analysis of counterfeit herbal drugs (12-14,16,24,60,62). These two methods have core similarities in that both perform drug analysis on a chromatographic basis. However, HPLC and GC/MS operate on entirely different mechanisms. Key differences in these two methods are: (1) mobile and stationary phases, and (2) separation mechanisms. Both methods are used as efficient techniques for the analysis of hidden APIs in counterfeit drugs in forensic toxicology laboratories (24).

4.6   NECESSITY FOR METHOD VALIDATION IN COUNTERFEIT DRUG ANALYSIS

Results obtained from toxicological analysis should be reliable, provide trustworthy information, and be able to analyze a sample containing analytes for well-defined purposes. To plan for the development of a toxicological analysis, some points – such as the purpose of analysis, type of analyte, the nature of the sample or matrix, and the analytical method – should be considered to ensure accurate, reliable, and reproducible results (63).

4.6.1 METHOD VALIDATION

Simultaneous determination of different analytes in experiment matrices should be fully validated. It is necessary to evaluate some parameters, such as: specificity, precision and accuracy, limit of detection (LOD) and limit of quantification (LOQ), and linearity prior to routine analysis of samples (63).

4.6.1.1 Selectivity

Selectivity of the proposed method is defined as the method that can determine a particular analyte in the matrix without interfering with other drugs and substances in the matrix medium. Selectivity is evaluated by comparing typical chromatograms obtained by analyzing a blank sample (mobile phase without analyte), solution-containing standard and a standard, validation solution (standard and other interfering substances). Results should indicate the absence of peaks at the retention time corresponding to the peak of analyte of interest (63).

4.6.1.2 Accuracy

Accuracy of a method is defined as the agreement between the results of an analytical method and the true concentration of the analyte in the sample. In other words, accuracy evaluates the systematic error of an analytical method.

4.6.1.3 Precision

Random errors are calculated as precision or may be expressed as reproducibility.

4.6.1.4 Limit of Detection (LOD)

Limit of detection is the lowest concentration of an analyte that can be detected in a sample.

4.6.1.5 Limit of Quantification (LOQ)

Limit of quantification is the smallest amount of the analyte analyzed under the experimental condition.

4.6.1.6 Linearity

When the obtained test results are proportional to the concentration of an analyte in the experiment medium, then the quantitative method is linear.

4.6.1.7 Recovery of Extraction Efficiency

Recovery is the percentage of an analyte that moves into the extracting phase. Different sample preparation techniques do not always provide the same recoveries for an analyte (63).

For example, Deconinck *et al.*, developed and validated a fast headspace GC/MS method for the quantitative analysis of residual solvents in counterfeit tablets and capsules (64). Linearity of the calibration line, selectivity, precision, accuracy and uncertainty, LOD, LOQ, and recovery were obtained for ethanol, 2-propanol, acetone, ethylacetate, dichloromethane and other solvents (64). Likewise, LC/MS method was validated for the simultaneous determination of glucocorticoids in pharmaceutical formulations and counterfeit cosmetic products. The researchers concluded that steroids may unfortunately be added illegally to a large number of cosmetic products (53). Antibiotics purchased from informal markets were categorized as counterfeit and substandard drugs using a validated LC method. Linearity, mean recovery, and robustness of uncertainty were defined for amoxicillin, ampicillin, chloramphenicol, clavulanic acid, doxycycline, and cloxacillin (65). Similarly, UHPLC was validated and developed for the detection and quantification of erectile dysfunction drugs and their analogues in counterfeit medicines (15).

4.7 CONCLUSION

One of the most common applications of GC/MS is drug identification and quantification for clinical and forensic purposes. Hidden APIs in counterfeit herbal supplements can be identified by matching the full scan spectrum of obtained peaks from a GC/MS with a mass spectral library or a database. Mass spectrometers coupled with gas chromatographs serve as a versatile analytical technique that incorporates the separation power of gas chromatography with the specificity of a mass spectrometer. One limitation to GC/MS is the requirement that the analyte be sufficiently volatile to be transferred to the mobile phase and subsequently eluted from analytical column to the detector. However, GC/MS has many conclusive attributes that makes it one of the most versatile and beneficial analytical instruments with high efficiency chromatographic separation and good limits of quantification of drugs using spectral libraries. It also offers high sensitivity and specificity for unknown drugs analysis in biological and non-biological matrices.

REFERENCES

1. Ikhsan Arif Nur, Syifa Fella, Mustafidah Mabrurotul, and Rohman Abdul. "Implementation of chemometrics as a solution to detecting and preventing falsification of herbal medicines in Southeast Asia: a review." *Journal of Applied Pharmaceutical Science* 11, no. 09 (2021): 139–148. doi:10.7324/JAPS.2021.110917.
2. World Health Organization (WHO); *Counterfeit drugs—Guidelines for the development of measures to combat counterfeit drugs.* WHO/EDM/QSM/99.1. Geneva, Switzerland, (1999).

3. Newton Paul N, Green Michael D, Fernández Facundo M, Day Nicholas PJ, and White Nicholas J. "Counterfeit anti-infective drugs." *The Lancet Infectious Diseases* 6, no. 9 (2006): 602–613. doi:10.1016/S1473-3099(06)70581-3.

4. World Health Organization. Definitions of Substandard and Falsified (SF) Medical Products.   www.who.int/medicines/regulation/ssffc/definitions/en/   (accessed 2021-07-09).

5. Tobolkina Elena, and Rudaz Serge. "Capillary electrophoresis instruments for medical applications and falsified drug analysis/quality control in developing countries." *Analytical Chemistry* 15, no. 93 (2021): 8107–8115. doi:10.1021/acs.analchem.1c00839.

6. Bonsu, Dan Osei Mensah, Afoakwah Constance, and Aguilar-Caballos Maria de la Paz. "Counterfeit formulations: analytical perspective on anorectics." *Forensic Toxicology* (2021): 1–25. doi:org/10.1007/s11419-020-00564-5.

7. Dégardin Klara, Roggo Yves, and Margot Pierre. "Forensic intelligence for medicine anti-counterfeiting." *Forensic Science International* 248, (2015): 15–32. doi:org/10.1016/j.forsc iint.2014.11.015.

8. Bolla Anmole S, Patel Ashwani R, and Priefer Ronny. "The silent development of counterfeit medications in developing countries – a systematic review of detection technologies." *International Journal of Pharmaceutics* 587, (2020): 119702. doi:10.1016/j.ijpharm.2020.119702.

9. Jairoun Ammar A, Al-Hemyari Sabaa Saleh, Moyad Shahwan, and Zyoud Sa'ed H. "Adulteration of weight loss supplements by the illegal addition of synthetic pharmaceuticals." *Molecules* 26, no. 22 (2021): 6903. doi:10.3390/molecules26226903.

10. Etil Ariburnu, Uludag Mehmet Fazli, Yalcinkaya Huseyin, and Yesilada Erdem. "Comparative determination of sibutramine as an adulterant in natural slimming products by HPLC and HPTLC densitometry." *Journal of Pharmaceutical and Biomedical Analysis* 64 –65, (2012): 77–81.

11. Marjan Khazan, Mehdi Hedayati, Farzad Kobarfard, Sahar Askari, and Fereidoun Azizi. "Identification and determination of synthetic pharmaceuticals as adulterants in eight common herbal weight loss supplements." *Iranian Red Crescent Medical Journal* 16, no. 3 (2014): e15344. doi:10.5812/ircmj.15344.

12. Dastjerdi Ghasemi Akram, Akhgari Maryam, Kamali Artin, and Mousavi Zahra. "Principal component analysis of synthetic adulterants in herbal supplements advertised as weight loss drugs." *Complementary Therapies in Clinical Practice* 31 (2018): 236–241. doi:10.1016/j.ctcp.2018.03.007.

13. Foroughi Mohammad Hadi, Akhgari Maryam, Jokar Farzaneh, and Mousavi Zahra. "Identification of undeclared active pharmaceutical ingredients in counterfeit herbal medicines used as opioid substitution therapy." *Australian Journal of Forensic Sciences* 49, no. 6 (2017): 720–729. doi:org/10.1080/00450618.2016.1273387.

14. Fard, Hengameh Hafizi, and Akhgari Maryam. "Analytical perspectives of chemical adulterants in herbal sexual enhancer drugs." *Journal of Pharmacy & Pharmacognosy Research* 6, no. 1 (2018): 45–53. http://jppres.com/jppres.

15. Sacré Pierre-Yves, Deconinck Eric, Chiap Patrice, Crommen Jacques, Mansion François, Rozet Eric, Courselle Patricia, and De Beer Jacques O. "Development and validation of a ultra-high-performance liquid chromatography-UV method for the detection and quantification of erectile dysfunction drugs and some of their

analogues found in counterfeit medicines." *Journal of Chromatography A* 1218, no. 37 (2011): 6439–6447. doi:10.1016/j.chroma.2011.07.029.

16. Saberi Niosha, Akhgari Maryam, Bahmanabadi Leila, Bazmi Elham, and Mousavi Zahra. "Determination of synthetic pharmaceutical adulterants in herbal weight gain supplements sold in herb shops, Tehran, Iran." *DARU Journal of Pharmaceutical Sciences* 26, no. 2 (2018): 117–127. doi:10.1007/s40199-018-0216-2.

17. Alghannam, Abdulaziz, Aslanpour Zoe, Evans Sara, and Schifano Fabrizio. "A systematic review of counterfeit and substandard medicines in field quality surveys." *Integrated Pharmacy Research and Practice* 3, (2014): 71–88. dx.doi:org/10.2147/IPRP.S63690.

18. Martino Rosemary, Malet-Martino Myriam, Gilard Véronique, and Balayssac Stéphane. "Counterfeit drugs: analytical techniques for their identification." *Analytical and Bioanalytical Chemistry* 398, no. 1 (2010): 77–92. doi:10.1007/s00216-010-3748-y.

19. Benchekroun Yassine Hameda, El Karbane Miloud, Ihssane Bouchaib, Haidara Hasnaa, Azougagh Mohamed, and Saffaj Taoufiq. "Application of design space, uncertainty, and risk profile strategies to the development and validation of UPLC method for the characterization of four authorized phosphodiesterase type 5 inhibitors to combat counterfeit drugs." *Journal of AOAC International* 103, no. 3 (2020): 715–724. doi:10.1093/jaocint/qsz006. PMID: 33241372.

20. Hübschmann Hans-Joachim. *Handbook of GC-MS: fundamentals and applications* (Weinheim, Germany: John Wiley & Sons, 2015).

21. Lorenzo Maria, and Pico Yolanda, "Gas chromatography and mass spectroscopy techniques for the detection of chemical contaminants and residues in foods." in *Chemical Contaminants and Residues in Food*, ed. D Schrenk. (Woodhead Publishing, 2017), 15–50.

22. Yi He, and Concheiro-Guisan Marta. "Microextraction sample preparation techniques in forensic analytical toxicology." *Biomedical Chromatography* 33, no. 1 (2019): e4444. doi:org/10.1002/bmc.4444.

23. Zhang Qing-Wen, Lin Li-Gen, and Ye Wen-Cai. "Techniques for extraction and isolation of natural products: a comprehensive review." *Chinese Medicine* 13, no. 1 (2018): 1–26. doi:10.1186/s13020-018-0177-x.

24. Chen Xinlv, Wu Xinyan, Luan Tiangang, Jiang Ruifen, and Ouyang Gangfeng. "Sample preparation and instrumental methods for illicit drugs in environmental and biological samples: A review." *Journal of Chromatography A* 1640 (2021): 461961. doi:org/10.1016/j.chroma.2021.461961.

25. Wu Ling, Sun Rui, Li Yongxin, and Sun Chengjun. "Sample preparation and analytical methods for polycyclic aromatic hydrocarbons in sediment." *Trends in Environmental Analytical Chemistry* 24 (2019): e00074. doi.org/10.1016/j.teac.2019.e00074.

26. Mustafa Arwa, and Turner Charlotta. "Pressurized liquid extraction as a green approach in food and herbal plants extraction: A review." *Analytica Chimica Acta* 703, no. 1 (2011): 8–18. doi:10.1016/j.aca.2011.07.018.

27. Ahmadi-Jouibari Toraj, Fattahi Nazir, Shamsipur Mojtaba, and Pirsaheb Meghdad. "Dispersive liquid–liquid microextraction followed by high-performance liquid chromatography–ultraviolet detection to determination of opium alkaloids in human plasma." *Journal of Pharmaceutical and Biomedical Analysis* 85 (2013): 14–20. doi:10.1016/j.jpba.2013.06.030.

28. Vinatoru Mircea, Mason Timothy J, and Calinescu Ioan. "Ultrasonically assisted extraction (UAE) and microwave assisted extraction (MAE) of functional compounds from plant materials." *TrAC Trends in Analytical Chemistry* 97 (2017): 159–178. doi:org/10.1016/j.trac.2017.09.002.

29. Eskilsson Cecilia Sparr, and Björklund Erland. "Analytical-scale microwave-assisted extraction." *Journal of Chromatography A* 902, no. 1 (2000): 227–250. doi:10.1016/s0021-9673(00)00921-3.

30. Baciu Tatiana, Borrull Francesc, Aguilar Carme, and Calull Marta. "Findings in the hair of drug abusers using pressurized liquid extraction and solid-phase extraction coupled in-line with capillary electrophoresis." *Journal of Pharmaceutical and Biomedical Analysis* 131 (2016): 420–428. doi:10.1016/j.jpba.2016.09.017.

31. Pedersen-Bjergaard Stig, and Rasmussen Knut Einar. "Liquid-phase microextraction with porous hollow fibers, a miniaturized and highly flexible format for liquid–liquid extraction." *Journal of Chromatography A* 1184, no. 1-2 (2008): 132–142. doi:10.1016/j.chroma.2007.08.088.

32. Qriouet Zidane, Qmichou Zineb, Bouchoutrouch Nadia, Mahi Hassan, Cherrah Yahia, and Sefrioui Hassan. "Analytical methods used for the detection and quantification of benzodiazepines." *Journal of Analytical Methods in Chemistry* 2019 (2019): 2035492. doi:10.1155/2019/2035492.

33. Hennion Marie-Claire. "Solid-phase extraction: method development, sorbents, and coupling with liquid chromatography." *Journal of Chromatography A* 856, no. 1-2 (1999): 3–54. doi:10.1016/s0021-9673(99)00832-8.

34. Nadal Joan Carles, Anderson Kimberley L, Dargo Stuart, Joas Irvin, Salas Daniela, Borrull Francesc, Cormack Peter AG, Marcé Rosa Maria, and Fontanals Núria. "Microporous polymer microspheres with amphoteric character for the solid-phase extraction of acidic and basic analytes." *Journal of Chromatography A* 1626 (2020): 461348. doi:10.1016/j.chroma.2020.461348.

35. Ettore Ferrari Júnior, and Caldas Eloisa Dutra. "Simultaneous determination of drugs and pesticides in postmortem blood using dispersive solid-phase extraction and large volume injection-programmed temperature vaporization-gas chromatography–mass spectrometry." *Forensic Science International* 290 (2018): 318–326. doi:10.1016/j.forsciint.2018.07.031.

36. Rodrigues Taís B, Morais Damila R, Gianvecchio Victor AP, Aquino Elvis M, Cunha Ricardo L, Huestis Marilyn A, and Costa Jose Luiz. "Development and validation of a method for quantification of 28 psychotropic drugs in postmortem blood samples by modified Micro-QuEChERS and LC–MS-MS." *Journal of Analytical Toxicology* 45, no. 7 (2021): 644–656. doi:10.1093/jat/bkaa138.

37. Kim Beomhee, Lee Wonwoong, Kim Youlee, Lee Jihyun, and Hong Jongki. "Efficient matrix cleanup of soft-gel-type dietary supplements for rapid screening of 92 illegal adulterants using EMR-lipid dSPE and UHPLC-Q/TOF-MS." *Pharmaceuticals* 14, no. 6 (2021): 570. doi:10.3390/ph14060570.

38. Deconinck Eric, Sacré Pierre-Yves, Courselle Patricia, and De Beer Jacques O. "Chromatography in the detection and characterization of illegal pharmaceutical preparations." *Journal of Chromatographic Science* 51, no. 8 (2013): 791–806. doi:10.1093/chromsci/bmt006.

39. Lu Feng, Weng Xinxin, Chai Yifeng, Yang Yongjian, Yu Yinjia, and Duan Gengli. "A novel identification system for counterfeit drugs based on portable Raman

spectroscopy." *Chemometrics and Intelligent Laboratory Systems* 127 (2013): 63–69. doi:10.1016/j.chemolab.2013.06.001.

40. Anzanello Michel J, Fogliatto Flavio S, Ortiz Rafael S, Limberger Renata, and Mariotti Kristiane. "Selecting relevant Fourier transform infrared spectroscopy wavenumbers for clustering authentic and counterfeit drug samples." *Science & Justice* 54, no. 5 (2014): 363–368. doi:10.1016/j.scijus.2014.04.005.

41. Custers Deborah, Canfyn Michaël, Courselle Patricia, De Beer Jacques Omer, Apers Sandra, and Deconinck Eric. "Headspace–gas chromatographic fingerprints to discriminate and classify counterfeit medicines." *Talanta* 123 (2014): 78–88. doi:org/10.1016/j.talanta.2014.01.020.

42. Bunaciu Andrei A, and Aboul-Enein Hassan Y. "Adulterated drug analysis using FTIR spectroscopy." *Applied Spectroscopy Reviews* 56, no. 5 (2021): 423–437. doi:org/10.1080/05704928.2020.1811717.

43. Oberacher Herbert, and Arnhard Kathrin. "Current status of non-targeted liquid chromatography-tandem mass spectrometry in forensic toxicology." *TrAC Trends in Analytical Chemistry* 84 (2016): 94–105. doi:org/10.1016/j.trac.2015.12.019.

44. Pragst Fritz, "High performance liquid chromatography in forensic toxicological analysis," in: *Handbook of Analytical Separations*, ed. Bogusz, Macej J (Elsevier, 2008), 447–489. doi:org/10.1016/S1567-7192(06)06013-X.

45. Rawtani Deepak, Khatri Nitasha, Tyagi Sanjiv, and Pandey Gaurav. "Nanotechnology-based recent approaches for sensing and remediation of pesticides." *Journal of Environmental Management* 206 (2018): 749–762. doi:10.1016/j.jenvman.2017.11.037.

46. Carlin Michelle Groves, and Dean John Richard, *Forensic Applications of Gas Chromatography*, (Boca Raton: CRC Press, 2013).

47. United States Office of Forensic Sciences, Microgram bulletin, Drug Enforcement Administration, December 18, 2021. https://catalogue.nla.gov.au/Record/3832998.

48. Soltaninejad Kambiz, Faryadi Mansoor, Akhgari Maryam, and Bahmanabadi Leila. "Chemical profile of counterfeited buprenorphine vials seized in Tehran, Iran." *Forensic Science International* 172, no. 2-3 (2007): e4-e5. doi:10.1016/j.forsciint.2007.06.016.

49. Committee on Understanding the Global Public Health Implications of Substandard, Falsified, and Counterfeit Medical Products; Board on Global Health; Institute of Medicine; Gillian J Buckley and Lawrence O Gostin eds. *Countering the Problem of Falsified and Substandard Drugs*. Washington (DC): National Academies Press (US); 2013 May 20.

50. Fabresse Nicolas, Gheddar Laurie, Kintz Pascal, Knapp Adeline, Larabi Islam Amine and Alvarez Jean-Claude. "Analysis of pharmaceutical products and dietary supplements seized from the black market among bodybuilders." *Forensic Science International* 322 (2021): 110771. doi:10.1016/j.forsciint.2021.110771.

51. Rockwood Alan L, Kushnir Mark M, and Clarke Nigel J. "Mass spectrometry." In: *Principles and Applications of Clinical Mass Spectrometry, Small Molecules, Peptides, and Pathogens*, Rifai Nader, Horvath Andrea Rita and Wittwer Carl T eds. (Elsevier, 2018), 33–65. doi:org/10.1016/B978-0-12-816063-3.00002-5.

52. Lebel Philippe, Gagnon Jacques, Furtos Alexandra and Waldron Karen C. "A rapid, quantitative liquid chromatography-mass spectrometry screening method for 71 active and 11 natural erectile dysfunction ingredients present in potentially adulterated or counterfeit products." *Journal of chromatography A* 1343

(2014): 143–151. doi:10.1016/j.chroma.2014.03.078https://doi.org/10.1016/j.chroma.2014.03.078

53. Fiori Jessica, and Andrisano Vincenza. "LC–MS method for the simultaneous determination of six glucocorticoids in pharmaceutical formulations and counterfeit cosmetic products." *Journal of Pharmaceutical and Biomedical Analysis* 91 (2014): 185–192. doi:10.1016/j.jpba.2013.12.026.

54. Ortiz Rafael S, Mariotti Kristiane de Cássia, Fank Bruna, Limberger Renata P, Anzanello Michel J, and Mayorga Paulo. "Counterfeit Cialis and Viagra fingerprinting by ATR-FTIR spectroscopy with chemometry: can the same pharmaceutical powder mixture be used to falsify two medicines?." *Forensic Science International* 226, no. 1-3 (2013): 282–289. doi:10.1016/j.forsciint.2013.01.043.

55. Cho So-Hyun, Park Hyoung Joon, Lee Ji Hyun, Do Jung-Ah, Heo Seok, Jo Jeong Hwa, and Cho Sooyeul. "Determination of anabolic–androgenic steroid adulterants in counterfeit drugs by UHPLC–MS/MS." *Journal of Pharmaceutical and Biomedical Analysis* 111 (2015): 138–146. doi:10.1016/j.jpba.2015.03.018.

56. United States Office of Forensic Sciences, Microgram bulletin, Drug Enforcement Administration. https://catalogue.nla.gov.au/Record/3832998.

57. Liu Song-Yun, Woo Soo-On, and Koh Hwee-Ling. "HPLC and GC–MS screening of Chinese proprietary medicine for undeclared therapeutic substances." *Journal of Pharmaceutical and Biomedical Analysis* 24, no. 5-6 (2001): 983–992. doi:10.1016/s0731-7085(00)00571-9.

58. Musshoff Frank, Daldrup Thomas, and Ritsch M. "Anabolic steroids on the German black market." *Archiv Fur Kriminologie* 199, no. 5-6 (1997): 152–158.

59. Alabdalla Mahmoud A. "Chemical characterization of counterfeit captagon tablets seized in Jordan." *Forensic Science International* 152, no. 2-3 (2005): 185–188. doi:10.1016/j.forsciint.2004.08.004.

60. Akhgari Maryam, Mohammadi Bahman Haj, Jokar Farzaneh, Mousavi Zahra. "Determining the effective substance of prevalent super slim weight loss capsule." *Asia Pacific Journal of Medical Toxicology* 7, no. 4 (2018): 100–106.

61. Da Justa Neves Diana Brito, and Caldas Eloisa Dutra. "GC–MS quantitative analysis of black market pharmaceutical products containing anabolic androgenic steroids seized by the Brazilian Federal Police." *Forensic Science International* 275 (2017): 272–281. doi:10.1016/j.forsciint.2017.03.016.

62. Hachem Rabab, Assemat Gaëtan, Martins Nathalie, Balayssac Stéphane, Gilard Véronique, Martino Robert, and Malet-Martino Myriam. "Proton NMR for detection, identification and quantification of adulterants in 160 herbal food supplements marketed for weight loss." *Journal of Pharmaceutical and Biomedical Analysis* 124 (2016): 34–47. doi:10.1016/j.jpba.2016.02.022.

63. Benchekroun Yassine Hameda, El Karbane Miloud, Ihssane Bouchaib, Haidara Hasnaa, Azougagh Mohamed, and Saffaj Taoufiq. "Application of design space, uncertainty, and risk profile strategies to the development and validation of UPLC method for the characterization of four authorized phosphodiesterase type 5 inhibitors to combat counterfeit drugs." *Journal of AOAC International* 103, no. 3 (2020): 715–724. doi:10.1093/jaocint/qsz006.

64. Deconinck Eric, Canfyn Michaël, Sacré P-Y., Baudewyns Sébastien, Courselle Patricia, and De Beer Jacques O. "A validated GC–MS method for the determination and quantification of residual solvents in counterfeit tablets and capsules." *Journal*

of Pharmaceutical and Biomedical Analysis 70 (2012): 64–70. doi:10.1016/j.jpba.2012.05.022.

65. Gaudiano Maria Cristina, Di Maggio Anna, Antoniella Eleonora, Valvo Luisa, Bertocchi Paola, Manna Livia, Bartolomei Monica, Alimonti Stefano, and Rodomonte Andrea Luca. "An LC method for the simultaneous screening of some common counterfeit and sub-standard antibiotics: validation and uncertainty estimation." *Journal of Pharmaceutical and Biomedical Analysis* 48, no. 2 (2008): 303–309. doi:10.1016/j.jpba.2007.12.032.

5 Applications of UV-Vis Spectroscopy in Counterfeit Medications

Jin Ba¹ and Ronny Priefer¹*
¹MCPHS University, 179 Longwood Ave, Boston, MA, USA, 02115

CONTENTS

5.1 Introduction 119
5.2 Ultraviolet Visible Spectroscopy 120
5.3 Quantitative UV-Vis 121
5.4 Qualitative UV-Vis 124
5.5 Conclusion...125
References...126

5.1 INTRODUCTION

Safety regulations and quality control are crucial aspects to medication development. Unfortunately, in developing countries drug quality control measures are often not as stringent as in developed countries [1]. The circulation of these poor-quality drugs impact both pharmaceutical companies and patient safety. This is global public health crisis and effects both a country's economy and the residents' well-being. Medication theft is just one part in the circulation of counterfeit drugs [2]. By stealing medications that are already approved or still in clinical trials, companies can mimic these and sell the counterfeit drugs. The development of counterfeit drugs does not require money to be spent on research, which is the largest cost related to bringing a drug onto the market. Counterfeit drugs are significantly lower in out-of-pocket expense than validated medications, making them attractive for purchasing. Many online pharmacies are illegally selling counterfeits, with some being sold to the patient prior to the official approval of the genuine medication [2]. Medication theft has led to more than €10 billion lost revenue each year by pharmaceutical companies and an additional €1.7 billion by various governments [3]. Moreover, the manufacturing practices of counterfeit drugs typically fall short of the standards of drug production. Non-sterile environments result in poor-quality drugs that can be harmful and potentially lead to therapy failure [4]. Wrong doses

DOI: 10.1201/9781003270461-5

of medications coupled with poor quality, are also common in counterfeit drugs [5]. Due to insufficient active ingredients in some of these medications, counterfeit drug users do not achieve the desired therapeutic effects. Negative impacts by counterfeits include drug over/under dosing, delayed efficacy of medication, and toxic side effects [6].

Additionally, the trend of counterfeit opioids is also increasing [7]. A major component in these counterfeit opioids is fentanyl. Some alprazolam and oxycodone tablets have been reported to contain fentanyl, which can cause involuntary opioid overdose, leading to life-threatening respiratory depression [7]. Likewise, an increasing amount of counterfeit COVID-19 vaccines have been detected on the market, which inadvertently led to a speed up in the transmission of the coronavirus [8]. Several cases of seizures in Africa and South-East Asia were also reported as a direct consequence of injecting falsified COVID-19 vaccines [9]. The large demand for online pharmacies during the COVID-19 pandemic also subsequently accelerated counterfeit drug production [8].

Counterfeit drugs can mimic the shape, size, color, and taste of authentic medication. Thus, the determination of drug quality cannot be solely relied upon by appearance. Insufficient funds and lack of well-trained staff in developing countries are significant challenges for tackling the detection of counterfeits [10]. Conducting drug quality control and analysis by using advanced technologies in developing countries (and some developed countries) is crucial to prevent the circulation of counterfeit drugs; however, the cost is often an insurmountable hurdle.

Current techniques which include, nuclear magnetic resonance (NMR) spectroscopy, thin-layer chromatography (TLC), high performance liquid chromatography (HPLC), mass spectrometry (MS), and vibrational spectroscopies (Ramen or IR) have been used in detection and drug analysis [10]. Ultraviolet-visible spectroscopy (UV-Vis) is a relatively more economical technology that has been utilized. Additionally, a key benefit with UV-Vis is that, as of late, the cost for this technology has shown a significant reduction [2].

5.2   ULTRAVIOLET VISIBLE SPECTROSCOPY

UV-Vis is an analytical technique employed in both industry as well as regulatory agencies, such as the US Food and Drug Administration (FDA). The applications of UV-Vis in the pharmaceutical industry include drug development and quality control. UV-Vis measures and compares the amount of absorbed light by a test sample compared to a reference sample. Wavelengths being scanned typically range from 200 nm to 900 nm [11]. It is a type of absorption spectroscopy, whereby compounds absorb light energy in the ultraviolet visible range, which can subsequently be recorded. The spectra for each compound or mixture is unique, thus making comparison relatively easy. Counterfeit drugs may have very similar spectra to an authentic medication, but at different concentrations; thus, potentially being sub-therapeutic [11].

UV-Vis can be utilized either quantitatively or qualitatively to analyze target materials. Qualitative analysis is designed for recognizing the presence of a specific substance. Conversely, the actual amount of a substance is determined by quantitative analysis. To test the purity of a compound, it is possible to measure the absorbance at a specific UV-Vis wavelength. If the analyte contains impurities, the absorbance at that wavelength may differ from a genuine sample. Using quantitative analysis, Beer-Lambert's Law is applied to determine the amount of the active ingredient, since the concentration of the compound of interest is directly proportional to the absorbed light of the sample [12]. Unfortunately, several factors may affect the accuracy of UV-Vis analysis, such as: temperature, solution pH, dilution, and so forth. [12].

Recently, Clopidogrel (Plavix®), an anti-platelet medication, was found on the market in a falsified form utilizing UV-Vis spectroscopy [13]. The amount of clopidogrel was directly determined by measuring its absorbance at its λ_{max}. By comparing this absorbance to an authentic sample of clopidogrel, the accurate concentration was elucidated [13].

Another example compared the isolated unknown impurities in levothyroxine with known inactive ingredients of an authentic formulation [14]. It was reported that acetonitrile was present in higher amounts in the counterfeit. The counterfeit medication not only lacked the active pharmaceutical ingredient but also contained some unknown inactive ingredients [14]. These unknown components could be harmful; thus, the detection and analysis of all ingredients are necessary.

Beyond small molecule drugs, biologics require unique purification, which can be validated by UV-Vis. After the separation of lysozyme from cytochrome c via column chromatography, it is possible to determine the levels of impurities of the final biologic medication by utilizing UV-Vis [15].

5.3 QUANTITATIVE UV-VIS

Mbinze *et al.,* employed UV-Vis to test the authenticity of amoxicillin, metronidazole, and quinine [16]. These medications are prescribed to treat malaria infection. Poor quality has given rise to antibiotic resistance in many Africa and Southeast Asia countries. To analyze the quality of these medications, analytical technology, such as UV-Vis, has been employed. Four samples – quinine sulfate, quinine bichlorhydrate, metronidazole benzoate, and amoxicillin trihydrate – were purchased in the Democratic Republic of the Congo (DRC). Likewise, four complimentary, authentic, reference formulations were obtained from Belgium, the United Kingdom, and Germany. To prepare samples for testing, the solid doses were dissolved and diluted to obtain solutions of different concentrations. These solutions were compared to the authentic formulation, which was dissolved in an identical manner. To test the accuracy of quantification by UV-Vis, the targeted amounts of the authentic formulations were 20% quinine in oral drops, metronidazole 125 mg in oral suspension, as well as 500 mg each of quinine and amoxicillin in tablets. [16].

Amoxicillin has a λ_{max} wavelength at 229 nm, while both metronidazole and quinine are at approximately 235 nm [16]. By comparing the absorbance from the DRC samples to those from pure formulations, the specific amount of quinine sulfate, quinine bichlorhydrate, metronidazole benzoate, and amoxicillin trihydrate was determined. Likewise, by comparing the absorbance of samples at its $\lambda_{max,}$ the accuracy of UV-Vis could be approved. For these, it was concluded that the DRC samples were accurate and within 10% acceptance limits.[16].

Mabrouk *et al.*, utilized UV-Vis to analyze and quantify sildenafil in counterfeits [17]. Sildenafil, a PDE-5 inhibitor, is involved in treating erectile dysfunction and pulmonary hypertension [18]. Counterfeits have been reported to contain sildenafil in a mixture with metronidazole, glyburide, and/or paracetamol. A combination of this PDE-5 inhibitor with one of these medications can cause significant drug interactions. Specifically, glyburide with sildenafil produces severe hypoglycemia, which can be fatal [19]. The researchers looked to compare counterfeits to an authentic sample; thus sildenafil, paracetamol, metronidazole, and glyburide purchased from Egypt were tested [17].

Identifying sildenafil in a mixture is not necessarily trivial because of overlapping spectra between it and the other potential additives. One approach employed involved scaling factors. Scaling factors are commonly used to adjust overlapping signals by increasing the amplitude of the spectra. The adjusted spectra with peak amplitude at zero-crossing wavelengths, can thus be recorded. [20]. Sildenafil in a mixture with paracetamol, has a λ_{max} absorbance at 311.8 nm, while paracetamol alone absorbs at 265.8 nm, with minimal to no cross-over. Similarly, in a metronidazole and sildenafil mixture, sildenafil has a λ_{max} at 319.5 nm, while metronidazole has a maximum absorbance at 291.8 nm, without overlap. Meanwhile, in a glyburide and sildenafil mixture, the latter display a λ_{max} at 328 nm, while glyburide alone has an absorbance at 316.3 nm, which the aforementioned scaling factor resolved [17].

Additionally, sildenafil has also been found in some herbal products. Pandey and Parikh compared extracted sildenafil citrate from herbal products to that found in genuine Viagra®[21]. Sildenafil citrate has a maximum absorbance at 292nm, thus a simple scan from 200–400 nm was done to confirm a λ_{max} match. Initially, extracts of the standard drug and different herbal preparations were done in a methanol–water mixture. It was observed that two of the seven herbal product samples were adulterated with sildenafil citrate, with as much as 520 µg/ml. In fact, the concentration of sildenafil citrate in both of the adulterated herbal products was higher than in the standard drug, which only contains 6 µg/ml [21].

Tilki *et al.*, determined the quality of paracetamol (acetaminophen) from different brands by utilizing UV-Vis [22]. Paracetamol, an analgesic, is used to treat mild or chronic pain [23]. Counterfeit paracetamol has been reportedly sold in several developing countries. Excessive doses of paracetamol can lead to fatal liver damage. Four brands of paracetamol tablets were purchased from Greater Noida and India. The standard of authenticity specifies no less than 95% and no more than 105%, paracetamol. Samples were dissolved and their absorbance at its λ_{max}, 257 nm, was recorded. Amounts of paracetamol in both authentic and samples were calculated by the proportion of weight and absorbances. The actual quantity of paracetamol in the tablets of the four brands were 0.48106 g, 0.49970 g, 0.50110 g, and 0.48725 g, which were comparable to authentic. Thus, these four brands were not deemed counterfeits [22].

Terazosin, a selective α1-blocker, is prescribed to treat benign prostatic hyperplasia (BPH) [24]. Conversely, prazosin is a non-selective α-blocker and utilized off-label for BPH [25]. Counterfeit tablets have been reported to contain both terazosin and prazosin in tablets. A combination of these two drugs can induce serious side effects. Ibrahim *et al.*, employed UV-Vis to differentiate prazosin from terazosin [26]. Samples were purchased from Egypt, while authentic terazosin was provided by Egypt's national organization. Terazosin has a λ_{max} at 254 nm while prazosin is at 273.9 nm. Due to overlapping spectra, machine learning models, and variable selection algorithms were applied to predict the concentrations of terazosin and prazosin. Comparing samples to the reference revealed that prazosin was indeed presented in the terazosin tablets. An algorithm model was used to predict the final concentration of terazosin. [26].

To control the quantity of ibuprofen and famotidine in a combined formulation, Nyola *et al.*, applied UV-Vis to predict the concentrations of these two medications in tablets [27]. Ibuprofen is a non-steroidal, anti-inflammatory medication that can be used to treat arthritis, fever, and pain. Ibuprofen has been indicated to increase the risk of ulceration. Famotidine, an H_2 blocker, thus can be used to reduce the amount of acid in the stomach. A combination of ibuprofen and famotidine can be prescribed to treat osteoarthritis and reduce ulceration caused by ibuprofen. Ibuprofen has a λ_{max} at 224 nm, while famotidine has the maximal absorbance at 286 nm. The standard concentrations of ibuprofen and famotidine are 12 µg/ml and 6 µg/ml, respectively. Samples of combined ibuprofen and famotidine, with a centration at 16 µg/ml and 6 µg/ml, respectively, were purchased in the UK. Ultimately, the results shows that ibuprofen concentration in the sample is higher than in the authentic drug [27].

Similarly, Hoang *et el.* utilized UV-Vis to analyze the concentrations of combination ibuprofen and paracetamol [28]. Combination of these two drugs is used to treat fever among pediatric patients. However, by adding paracetamol to ibuprofen, the risk of excessive dosing develops, thus requiring careful monitoring. In authentic formulation, ibuprofen has a λ_{max} at 276.5 nm with a concentration between 12–32 mg/L, while paracetamol is at 256.8 nm with a concentration between 20–40 mg/L. Commercial formulations of ibuprofen and paracetamol were purchased from Vietnam. The difference in maximal absorbance between the authentic and the sample indicate that the samples contain less paracetamol. However, the maximal absorbance and concentration of ibuprofen corresponded to the authentic sample [28].

Storage conditions, such as temperature and light may alter the quality of medications. Ramya *et al.*, utilized UV-Vis to examine the quality of lamotrigine under different storage conditions [29]. Lamotrigine, an anticonvulsant, is mainly used to treat seizures by inhibiting sodium channels and the release of glutamate [30]. Authentic lamotrigine, which has a λ_{max} at 304 nm, was purchased from India and used for spectral measurements. Lamotrigine was exposed to sunlight, freezing point, and a light-resistant environment. The mean absorbance values after exposure to these environments ended up to be 0.3537, 0.5538, and 0.5996, respectively. This suggested that lamotrigine degraded based on storage conditions, with sunlight having the most noticeable effect. [29].

5.4 QUALITATIVE UV-VIS

Mensah *et al.,* applied UV-Vis to detect expired antimalarial herbal medications on the market [31]. In Ghana, herbal products are commonly used to treat malaria. However, expired herbal medications are still often sold, lowering the effectiveness of these products. The quality of herbal medicinal products was evaluated by comparing the λ_{max} of an authentic form to the samples. A total of 167 samples purchased from Ghana, were analyzed by UV-Vis. The authentic herbal medicinal product had a λ_{max} of 375 nm, while the expired herbal medicinal product displayed a lower absorbance value at this wavelength. Ultimately, it was determined that two of 176 samples were expired [31].

Baratta *et al.,* employed UV-Vis to specifically look for counterfeit medications on the market [32]. A total of 221 samples were purchased from authorized and nonauthorized pharmacies in India. Samples included antibiotics, anti-inflammatories, antipyretics, and so forth. It was determined that half of the samples were counterfeit. Shockingly, 2% of the samples contained absolutely no active pharmaceutical ingredients. Most of the substandard medications were from street pharmacies [32].

It has been suggested most substandard medications are sold by online pharmacies. Gelatti *et al.,* applied UV-Vis to analyze the quality of fluoxetine sold online [33]. Fluoxetine (Prozac®) is used to treat depression. A total of 13 samples were purchased from different websites, with 10 samples being produced in India. One sample was from Turkey and the remaining two were from the UK. To determine the quality of the samples, the UV-Vis was compared to authentic Prozac. It was reported that all samples from these online stores were substandard [33].

Analysis of beverages and food products have also utilized UV-Vis. Ríos-Reina *et al.,* employed UV-Vis to discriminate wine vinegar from unauthentic versions [34]. To prevent adulteration and counterfeiting, factories must differentiate wine vinegars by production method, aging, and the protected denomination of origin (PDO). Wine vinegars have different ages, with older brands being more costly. Counterfeits on the market are often mislabeled as PDO-aged. To determine authentic versus counterfeits, four groups – i.e., PDO-aged, PDO without aging, non-PDO-aged, and non-PDO without aging – were analyzed by UV-Vis. PDO without aging wine vinegars had a λ_{max} at 300 nm, while PDO-aged wine vinegars were red-shifted to appear between 325 nm and 450 nm. Non-PDO aged wine vinegars display maximal peak absorbance between 290 nm and 400 nm. Unfortunately, non-PDO without aging vinegars appear at 300 nm, which mimics that of PDO without aging [34].

Additionally, whisky brands can also be differentiated by UV-Vis. Martins *et al.*, analyzed seven brands of whiskey [35]. Components, such as furfural, hydroxymethylfurfural, and congeners make certain brands distinguishable as they have different absorbances. For instance, beverages which contain furfural have an absorbance at 210 nm, while those with hydroxymethylfurfural have a λ_{max} at 282 nm. The addition of these ingredients can help differentiate whisky brands as well as identify counterfeits [35].

The quantity of food dyes used in products must also be carefully managed. Bişgin utilized UV-Vis to quantify two food dyes – brilliant blue (BB) and sunset yellow (SY) [36]. Food dyes are broadly used in the cosmetic, pharmaceutical, and food industries for modifying the appearance of colored products. The World Health Organization recommended daily intakes of BB and SY should not exceed 6mg/kg and 4mg/kg, respectively. SY has a λ_{max} at 483 nm, while BB is at 630 nm. Some food samples purchased from Turkish markets labeled as no BB or SY added still show the presence of these dyes through UV-Vis analysis. An energy drink and citrus syrup contained 5.76 µg/mL and 142.49 µg/mL of BB, respectively, while orange juice powder and sesame sugar contain 331.8 µg/g and 21.67 µg/g of SY, respectively [36].

5.5   CONCLUSION

UV-Vis spectroscopy has demonstrated its versatility in detecting and quantifying impurities in counterfeits. This can be done by simply comparing the maximum absorbance of counterfeits to authentic samples. Compared to other detecting techniques, such as HPLC, NMR, and so forth, UV-Vis is less time consuming and lower in cost [37]. Additionally, UV-Vis is relatively easy to operate [38]. For developing countries, complex and expensive techniques require a long period of training, which UV-Vis minimizes. Lack of staff and insufficient budgets makes it difficult to apply expensive techniques for counterfeit detection. Implementation of UV-Vis allows for qualitative and/or quantitative analysis. In addition to medications, beverages and food dyes have also been analyzed using this simple technique. It is, however, important to consider that the accuracy of measurements by UV-Vis is influenced by external factors, such as temperature and solvent [39, 40]. Additionally, overlapping signals can convolute the spectrum [41]. To solve this problem, algorithm models [26] and scaling factors [17] have been employed to decrease this effect.

UV-Vis has found applications beyond counterfeit drugs, such as in the arena of water quality control. Shi *et al.*, utilized this absorption technology to detect pollutants in water [42]. Drinking water quality is essential to public health, thus the presence of pollutants must be minimized. Drinking water collected from ground, lake, and surface sources were collected from Austria, Brazil, and

Ireland. Water quality is generally evaluated by four parameters: chemical oxygen demand (COD), total organic carbon (TOC), nitrate levels, as well as turbidity and dissolved organic carbon (DOC) [43]. Pollutants in water contain unique maximum absorbances which can be detected by UV-Vis [42]. Nitrates and TOC display λ_{max} between 200–385 nm, while the maximal absorbance of COD is 220–250 nm, and DOC is at 254 nm. Turbidity determination tends to be very broad, between 200–720 nm maximal absorbance. The COD in the ground and surface water have λ_{max} at 350 nm and 546 nm, respectively. In the samples collected, the absorbance values of these two peaks were higher than anticipated, thus indicating that the COD levels were elevated. Likewise, DOC in surface water had an absorbance noticeably higher than that of pure water [42]. The rapid comparison between water samples and pure water allows for quick determination of pollutants that may require further removal.

UV-Vis has also been applied in mixed/mixing products. As with all medications, intravenous (IV) injections also require quality control. Angheluta *et al.*, utilized UV-Vis to analyze carbetocin (active pharmaceutical ingredient) and L-methionine (excipient) solutions [44]. The IV injection product, PAMFUL®, is a mixture of these two chemicals. Carbetocin displays two peaks, one in the range of 190–250 nm and another 250–300 nm. Conversely, L-methionine has only one peak in the range of 190–250 nm. To better control the quality of a mixture, the two independent components' concentrations were carefully monitored by UV-Vis. Clinically using these formulations is possible by comparing the concentrations obtained by UV-Vis of carbetocin and L-methionine, before and after mixing. As a result, it is possible to determine if a mixed product can be clinically administered to a patient [44].

UV-Vis is an efficient and easily operated technology for determining and quantifying counterfeit medications, in addition to beverages and various food products. This broad application has allowed UV-Vis to be used within pharmaceutical, food, and beverage industries as well as in regulatory agencies. Its low cost and time savings makes this technology favorable to help address the global crisis of counterfeit drugs.

REFERENCES

1. Glass B. Counterfeit drugs and medical devices in developing countries. *Research and Reports in Tropical Medicine.* 2014;5:11–22. doi:10.2147/rrtm.s39354
2. Ludasi K, Sovány T, Laczkovich O, Hopp B, Smausz T, Regdon G. Unique Laser Coding technology to fight falsified medicines. *European Journal of Pharmaceutical Sciences.* 2018;123:1–9. doi:10.1016/j.ejps.2018.07.023
3. Wajsman N, Burgos AC, Davies C. The economic cost of IPR infringement in the pharmaceutical industry. *EUIPO report.* 2016. https://euipo.europa.eu/tunnel-web/secure/webdav/guest/document_library/observatory/resources/research-and-stud ies/ip_infringement/study9/pharmaceutical_sector_en.pdf
4. Davison M. On-dose and in-dose authentication. Pharmaceutical Anti-Counterfeiting. 2011;103 –112. doi:10.1002/9781118023679.ch14

5. Ziavrou KS, Noguera S, Boumba VA. Trends in counterfeit drugs and pharmaceuticals before and during COVID-19 pandemic. *Forensic Science International*. 2022;338:111382. doi:10.1016/j.forsciint.2022.111382

6. Rahman MS, Yoshida N, Tsuboi H, Tomizu N, Endo J, Miyu O, Akimoto Y, Kimura K.. The health consequences of falsified medicines – a study of the published literature. *Tropical Medicine & International Health*. 2018;23(12):1294–1303. doi:10.1111/tmi.13161

7. Nguyen L, Evans A, Frank G, Levitas M, Mennella A, Short LC. Genuine and counterfeit prescription pill surveillance in Washington, D.C. *Forensic Science International*. 2022;339:111414. doi:10.1016/j.forsciint.2022.111414

8. Amankwah-Amoah J. Covid-19 and counterfeit vaccines: global implications, new challenges and opportunities. *Health Policy and Technology*. 2022;11(2):100630. doi:10.1016/j.hlpt.2022.100630

9. Medical product alert N°5/2021: Falsified COVISHIELD vaccine (update). World Health Organization. www.who.int/news/item/31-08-2021-medical-prod uct-alert-n-5-2021-falsified-covishield-vaccine. Published Augusr 31, 2021. Accessed October 25, 2022.

10. Martino R, Malet-Martino M, Gilard V, Balayssac S. Counterfeit drugs: analytical techniques for their identification. *Analytical and Bioanalytical Chemistry*. 2010;398(1):77–92. doi:10.1007/s00216-010-3748-y

11. Kafle BP. Theory and instrumentation of absorption spectroscopy. Chemical Analysis and Material Characterization by Spectrophotometry. 2020:17–38. doi:10.1016/b978-0-12-814866-2.00002-6

12. Villamena FA. UV–vis absorption and chemiluminescence techniques. Reactive Species Detection in Biology. 2017;203–251. doi:10.1016/b978-0-12-420017-3.00006-2

13. Shergill RS, Kristova P, Patel BA. Detection of falsified clopidogrel in the presence of excipients using voltammetry. *Analytical Methods*. 2021;13(44):5335–5342. doi:10.1039/d1ay01602d

14. Ruggenthaler M, Grass J, Schuh W, Huber CG, Reischl RJ. Levothyroxine sodium revisited: a wholistic structural elucidation approach of new impurities via HPLC-HRMS/MS, on-line H/D exchange, NMR spectroscopy and chemical synthesis. *Journal of Pharmaceutical and Biomedical Analysis*. 2017;135:140–152. doi:10.1016/j.jpba.2016.12.002

15. Brestrich N, Rüdt M, Büchler D, Hubbuch J. Selective protein quantification for preparative chromatography using variable pathlength UV/vis spectroscopy and partial least squares regression. *Chemical Engineering Science*. 2018;176:157–164. doi:10.1016/j.ces.2017.10.030

16. Mbinze JK, Mpasi JN, Maghe E, Kobo S, Mwanda R, Mulumba G, Bolande JB, Bayebila, TM, Amani MB, Hubert P, Marini RD. Application of total error strategy in validation of affordable and accessible UV-visible spectrophotometric methods for quality control of poor medicines. *American Journal of Analytical Chemistry*. 2015;06(02):106–117. doi:10.4236/ajac.2015.62010

17. Mabrouk M, Hammad S, Soliman B, Kamal A. Analysis of counterfeit sildenafil by validated UV spectrophotometric methods. *Journal of Advanced Medical and Pharmaceutical Research*. 2021. doi:10.21608/jampr.2021.89542.1016

18. Label: VIAGRA (sildenafil citrate) tablets. U.S. Food and Drug Administration Website. www.accessdata.fda.gov/drugsatfda_docs/label/2014/20895s039s042lbl. pdf. Revised March 2014. Accessed October 28, 2022.
19. Verma RK, Kumar R, Sankhla MS. Toxic effects of sexual drug overdose: sildenafil (viagra). *ARC Journal of Forensic Science*. 2019;4(1):26–31. doi:10.20431/ 2456-0049.0401003
20. Attimarad M, Venugopala KN, Aldhubiab BE, Nair AB, SreeHarsha N, Pottathil S, Akrawi SH. Development of UV spectrophotometric procedures for determination of amlodipine and celecoxib in formulation: use of scaling factor to improve the sensitivity. *Journal of Spectroscopy*. 2019;2019:1–10. doi:10.1155/2019/8202160
21. Pandey A, Parikh P. Detection of sildenafil citrate from aphrodisiac herbal formulations. Int*ernational* J*ournal of* Pharm*aceutical* Sci*ences and* Res*earch*. 2015;6(9): 4080–85.
22. Udeh-Momoh C, Watermeyer T. Female specific risk factors for the development of alzheimer's disease neuropathology and cognitive impairment: call for a precision medicine approach. *Ageing Research Reviews*. 2021;71:101459. doi:10.1016/ j.arr.2021.101459
23. Center for Drug Evaluation and Research. Acetaminophen. U.S. Food and Drug Administration.   www.fda.gov/drugs/information-drug-class/acetaminophen. Published September 6, 2022. Accessed October 20, 2022.
24. Hytrin-terazosin hydrochloride tablet. U.S. Food and Drug Administration Website. www.accessdata.fda.gov/drugsatfda_docs/label/2009/019057s022lbl.pdf. Revised July 2009. Accessed October 28, 2022.
25. Gordon SG, Kittleson MD. Drugs used in the management of heart disease and cardiac arrhythmias. Small Animal Clinical Pharmacology. 2008;380–457. doi:10.1016/b978-070202858-8.50019-1
26. Ibrahim AM, Hendawy HAM, Hassan WS, Shalaby A, ElMasry MS. Determination of terazosin in the presence of prazosin: different state-of-the-art machine learning algorithms with UV spectroscopy. *Spectrochimica Acta Part A: Molecular and Biomolecular Spectroscopy*. 2020;236:118349. doi:10.1016/j.saa.2020.118349
27. Nyola N, Govinda S, Kumavat M, Kalra N, Singh G. Simultaneous estimation of famotidine and ibuprofen in pure and pharmaceutical dosage form by UV-vis spectroscopy. *International Research Journal of Pharmacy*. 2012;3(4):277–280.
28. Hoang VD, Ly DT, Tho NH, Minh Thi Nguyen H. UV spectrophotometric simultaneous determination of paracetamol and ibuprofen in combined tablets by derivative and wavelet transforms. *The Scientific World Journal*. 2014;2014:1–13. doi:10.1155/2014/313609
29. Ramya T, Gunasekaran S, Ramkumaar GR. Density functional theory, restricted Hartree – Fock simulations and FTIR, FT-Raman and UV–vis spectroscopic studies on lamotrigine. *Spectrochimica Acta Part A: Molecular and Biomolecular Spectroscopy*. 2013;114:277–283. doi:10.1016/j.saa.2013.05.057
30. LAMICTAL Label. U.S. Food and Drug Administration Website. www.accessdata. fda.gov/drugsatfda_docs/label/2015/020241s045s051lbl.pdf. Revised March 2015. Accessed October 28, 2022.
31. Mensah JN, Brobbey AA, Addotey JN, Ayensu I, Asare-Nkansah S, Opuni KF, Adutwum LA. Ultraviolet-visible spectroscopy and chemometric strategy enable the classification and detection of expired antimalarial herbal medicinal product in

Ghana. *International Journal of Analytical Chemistry*. 2021;2021:1–9. doi:10.1155/2021/5592217

32. Baratta F, Germano A, Brusa P. Diffusion of counterfeit drugs in developing countries and stability of galenics stored for months under different conditions of temperature and relative humidity. *Croatian Medical Journal*. 2012;53(2):173–184. doi:10.3325/cmj.2012.53.173

33. Gelatti U, Pedrazzani R, Marcantoni C, Mascaretti S, Repice C, Filippucci L, Zerbini I, Dal Grande M, Orizio G, Feretti D. 'You've got m@il: Fluoxetine coming soon!': Accessibility and quality of a prescription drug sold on the web. *International Journal of Drug Policy*. 2013;24(5):392–401. doi:10.1016/j.drugpo.2013.01.006

34. Ríos-Reina R, Azcarate SM, Camiña J, Callejón RM, Amigo JM. Application of hierarchical classification models and reliability estimation by bootstrapping, for authentication and discrimination of wine vinegars by UV–vis spectroscopy. *Chemometrics and Intelligent Laboratory Systems*. 2019;191:42–53. doi:10.1016/j.chemolab.2019.06.001

35. Martins AR, Talhavini M, Vieira ML, Zacca JJ, Braga JW. Discrimination of whisky brands and counterfeit identification by UV–VIS spectroscopy and multivariate data analysis. *Food Chemistry*. 2017;229:142–151. doi:10.1016/j.foodchem.2017.02.024

36. Bişgin AT. Simultaneous preconcentration and determination of Brilliant Blue and sunset yellow in foodstuffs by solid-phase extraction combined UV-vis spectrophotometry. *Journal of AOAC International*. 2018;101(6):1850–1856. doi:10.5740/jaoacint.18-0089

37. Li P, Qu J, He Y, Bo Z, Pei M. Global calibration model of UV-VIS spectroscopy for COD estimation in the effluent of rural sewage treatment facilities. *RSC Advances*. 2020;10(35):20691–20700. doi:10.1039/c9ra10732k

38. Nasrollahzadeh M, Momeni SS, Sajadi SM. Green synthesis of copper nanoparticles using plantago asiatica leaf extract and their application for the cyanation of aldehydes using $K_4FE(CN)_6$. *Journal of Colloid and Interface Science*. 2017;506:471–477. doi:10.1016/j.jcis.2017.07.072

39. Bard B, Martel S, Carrupt P-A. High throughput UV method for the estimation of thermodynamic solubility and the determination of the solubility in biorelevant media. *European Journal of Pharmaceutical Sciences*. 2008;33(3):230–240. doi:10.1016/j.ejps.2007.12.002

40. Hendel T, Wuithschick M, Kettemann F, Birnbaum A, Rademann K, Polte J. In situ determination of colloidal gold concentrations with UV–VIS spectroscopy: Limitations and perspectives. *Analytical Chemistry*. 2014;86(22):11115–11124. doi:10.1021/ac502053s

41. Nachabé R, Hendriks BH, Van der Voort M, Desjardins AE, Sterenborg HJ. Estimation of biological chromophores using diffuse optical spectroscopy: benefit of extending the UV-vis wavelength range to include 1000 to 1600 nm. *Biomedical Optics Express*. 2010;1(5):1432–1442. doi:10.1364/boe.1.001432

42. Shi Z, Chow CW, Fabris R, Liu J, Jin B. Applications of online UV-vis spectrophotometer for drinking water quality monitoring and process control: a review. *Sensors*. 2022;22(8):2987. doi:10.3390/s22082987

43. Guo Y, Liu C, Ye R, Duan Q. Advances on water quality detection by UV-vis spectroscopy. *Applied Sciences*. 2020;10(19):6874. doi:10.3390/app10196874

44. Angheluta A, Guizani S, Saunier J, Rönnback R. Application of chemometric modelling to UV-vis spectroscopy: development of simultaneous API and critical excipient assay in a liquid solution continuous flow. *Pharmaceutical Development and Technology*. 2020;25(8):919–929. doi:10.1080/10837450.2020.1770789

6 IR Spectroscopic Analytical Tools in the Fight Against Counterfeit Medicines

Sangeeta Tanna[1,] and Rachel Armitage[2]*
[1]Leicester School of Pharmacy, De Montfort University, Leicester, United Kingdom
[2]Faculty of Health and Life Sciences, De Montfort University, Leicester, United Kingdom

CONTENTS

6.1 Introduction 132
6.2 Fundamentals of Infrared Spectroscopy...135
    6.2.1 Near-Infrared Spectroscopy...135
    6.2.2 Mid-Infrared Spectroscopy..137
        6.2.2.1 Attenuated Total Reflectance Fourier Transform Infrared Spectroscopy 138
    6.2.3 Comparison of NIR and ATR-FTIR Spectroscopy.......................139
6.3 Chemometric Approaches used in Combination with IR Spectroscopy 140
    6.3.1 Data Pre-Processing 142
    6.3.2 Unsupervised Methods..142
        6.3.2.1 Principal Component Analysis....................................142
    6.3.3 Supervised Methods 143
        6.3.3.1 Linear Discriminant Analysis 143
        6.3.3.2 Soft Independent Modelling of Class Analogies 143
        6.3.3.3 Data Driven-Soft Independent Modelling of Class Analogies 144
        6.3.3.4 Partial Least Squares-Discriminant Analysis................144
        6.3.3.5 k-Nearest Neighbour...144
        6.3.3.6 Classification and Regression Tree 145
    6.3.4 Regression Methods 145
        6.3.4.1 Principal Component Regression................................145
        6.3.4.2 Partial Least Squares Regression 146

DOI: 10.1201/9781003270461-6

6.4   Applications of IR Techniques for the Detection of Counterfeit
       and Substandard Medicines...146
       6.4.1   Near-Infrared Spectroscopy Applications 147
       6.4.2   Mid-Infrared Spectroscopy Applications 157
6.5   Conclusion...160
6.6   In memoriam: Dr Graham Lawson 161
6.7   Acknowledgments 161
References...161

6.1   INTRODUCTION

Medicine quality is of paramount importance for the safety of patients and fundamental to the success of health interventions. Counterfeit and substandard medicines constitute a growing public health and patient safety problem worldwide, particularly in low-income countries (LIC) and low to middle-income countries (LMIC). The growth in international trade together with internet sales have placed significant pressure on the assurance of pharmaceutical supply chain integrity in high-income countries (HIC). Counterfeit (falsified) medicines are deliberately and fraudulently produced and labelled, with packaging that is often indistinguishable from the genuine products, making them difficult to identify without running detection tests on the contents of the pharmaceutical dosage form. Substandard medicines result from poor manufacturing and quality assurance processes, as well as inadequate storage conditions, and reach the public due to poor regulatory controls or weak pharmaceutical governance (Sammons and Choonara, 2017). On the market, substandard and counterfeit medicines claim to be something they are not. These poor-quality pharmaceutical products are rarely efficacious and can lead to disastrous health consequences, including treatment failure, serious adverse drug reactions, disability and even death (WHO, 2017; Ghanem 2019). Additionally, they lead to unintentional medication nonadherence (Tanna and Lawson, 2016), increase the risk of drug resistance, undermine the public's confidence in healthcare systems and add to national economic burdens (WHO, 2017; Ghanem 2019). According to the World Health Organization (WHO), one in ten medical products in LMIC are substandard or falsified (WHO, 2017) although more recent reports advocate that approximately 40%–70% of medicines being sold in Africa are thought to be counterfeit or substandard (Bolla et al., 2020; Koech et al., 2020; Mwai 2020). The risks to patients are also significantly increased when medicines are purchased from unregulated websites, social media platforms and smartphone applications (WHO, 2018). Furthermore, there is considerable evidence that the COVID-19 pandemic has heightened the trade of counterfeit medicines, especially in LIC and LMIC, and this public healthcare problem is predicted to get worse (Newton et al., 2020; Waffo Tchounga et al., 2021). This rise is attributed to the major disruption in pharmaceutical supply chains and regular testing protocols due to national lockdowns, which has led to an increase in demand for low-cost medicines and an open market place for counterfeit medicines (Tesfaye et al., 2020).

The absence of the active pharmaceutical ingredient (API) or incorrect amounts of API are the main problems identified with counterfeit and substandard medicines, although they may include pharmaceutical products with the wrong ingredient(s), toxic impurities, dissolution failure, and fake packaging (Almuzaini et al., 2013). Generic and branded medicines for communicable and life-threatening diseases such as HIV/AIDS, tuberculosis and malaria, as well as those for chronic diseases such as cardiovascular disease, diabetes mellitus and cancer, have become the prime target of such poor-quality medicines (Alghannam et al., 2014). In Africa, due to the high burden of infectious diseases, counterfeit and substandard antibiotics are also widespread in the market (Koech et al., 2020). Nearly one in five antimalarials circulating in LMIC are counterfeit or substandard (Mackey 2018; Beargie et al., 2019). According to the WHO, there has also been an increase in counterfeit medicines related to COVID-19, including antiviral medicines and chloroquine (Cooper et al., 2020). In HIC, lifestyle drugs used to treat erectile dysfunction and weight loss and anabolic products have been extensively targeted (Rebiere et al., 2017). In recent years, biotechnology drugs, including vaccines, have been reported to be counterfeit (Janvier et al., 2018; Jarret et al., 2020; Srivastava 2021). This public health problem is not limited to expensive medicines, as conterfeits of low-priced medicines can still make a profit for criminals as long as the sales volume are high (Lawson et al., 2018) – hence cutting prices for licensed medicines will not provide a solution to this rising healthcare problem.

Given the high humanistic and economic cost associated with counterfeit and substandard medicines globally, there is a pressing need for the routine surveillance of pharmaceutical products in the pharmaceutical supply chain and at border controls to determine the authenticity of the products. The early detection of counterfeit and substandard medicines in a country will reduce the risk of these poor-quality products being consumed by patients. A variety of technologies from analytical chemistry can be used for surveillance (Rebiere et al., 2017; Bolla et al., 2020; Bakker-'t Hart et al., 2021); however, it remains a major challenge in LIC and LMIC due to limited resources and the lack of infrastructure and trained personnel. Analytical technologies vary considerably in characteristics that impact on their suitability for routine surveillance of medicines in these low-resource countries. Historically, validated pharmacopoeia approved analytical methods or non-validated inhouse procedures when pharmacopeial methods do not exist, have been employed by LIC and LMIC medicines quality control laboratories for determining the authenticity of medicines. In using this medicines authentication approach in low-resource countries, there are often significant delays between collection of suspicious medicines and confirmation of their poor quality, with harm spreading unchecked in the interim (Vickers et al., 2018). Techniques such as high-performance liquid chromatography (HPLC) with ultraviolet (UV) detection are used, which are destructive, time-consuming due to complex sample preparation steps, and expensive because they require large volumes of expensive solvents. Additionally, analytical technology requires well-trained personnel and well-equipped laboratories, which are not readily available in LIC and LMIC (Lawson et al., 2018). Simple analytical techniques requiring little or no sample preparation prior to analysis help to speed up analysis time and, thus, would be apt for first-line analysis of counterfeit medicines (Kovacs et al., 2014). Cost-effectiveness and portability are important features to consider when selecting analytical techniques for screening counterfeit and substandard medicines. Cheaper methods (in terms of cost of production and maintenance) will make the techniques more accessible to a wide range of users globally and, therefore, facilitate quicker analysis at different points in the pharmaceutical supply chain. Portability of the analytical equipment is also important in screening for counterfeit and substandard medicines in order to ensure ease of use in the field or at the point of sale of the medicines. The speed of analysis, cheaper costs, portability, and simplicity of the technique employed in the screening of medicines will go a long way in facilitating in-field analysis of medicines especially in LIC and LMIC. Due to the many constraints faced by low-resource countries, not all suspicious medicines are quality tested. Hence, there is an urgent need to empower LIC and LMIC officials at all levels of the pharmaceutical supply chain with user-friendly, low-cost, robust, non-destructive, and handheld or portable screening devices for the rapid detection of poor-quality medicines. Infrared (IR) spectroscopy-based devices fitting these criteria (Wilson et al., 2017; Vickers et al., 2018) could transform LIC and LMIC healthcare systems' medicine supply chains, providing assurance to health services and consumers, while damaging the profitability of counterfeit drug syndicates. If a pharmaceutical product is deemed poor-quality based on an initial IR spectroscopy screening in the field, it can be subjected to further laboratory testing using a confirmatory method, such as ambient ionisation mass spectrometry, to provide information about its authenticity.

Mid-infrared (MIR) spectroscopy and near-infrared (NIR) spectroscopy are versatile vibrational spectroscopy techniques, which can be used on solid or liquid samples and have found widespread application over the past two decades for the rapid screening of medicines. With spectroscopic-based methods, the excitation of fundamental molecular rotations or vibrations gives rise to unique spectra for the individual samples. The fundamental principles of these spectroscopy techniques are explained in Section 6.2. These fast and cost-effective IR techniques, which do not require consumables when combined with a spectral library, can provide conclusive information about pharmaceutical samples, and are ideally suited for both qualitative and quantitative analyses. Qualitative tests demonstrate the presence or absence of a specific API or component, while quantitative tests are able to confirm the levels of the API(s) in the sample (Kovacs et al., 2014). Additionally, workflow-based approaches can be implemented with these IR spectroscopy devices to enable inexperienced users to perform the analysis. When combined with classification chemometric algorithm(s), these MIR and NIR spectroscopy techniques can be used to verify the identity of a sample and detect counterfeit/substandard or suspect medicines. The chemometric tools employed with qualitative and quantitative MIR and NIR spectroscopy analyses are detailed in Section 6.3. Advances in sampling techniques, chemometric data analysis tools, ruggedness, and MIR and NIR spectroscopy instrument portability have allowed easy analysis of a broad range of pharmaceutical samples and the application of MIR and NIR spectroscopy to rapidly identify counterfeit and/or substandard medicines are reviewed and discussed in Section 6.4. In Section 6.5 the future directions for the application of IR spectroscopy devices for the rapid screening of counterfeit and substandard medicines are anticipated.

6.2 FUNDAMENTALS OF INFRARED SPECTROSCOPY

The measurement of absorption of IR radiation brought about by changes in molecular vibrations within molecules, gives rise to IR spectroscopy. The IR region of the electromagnetic spectrum encompasses radiation with wavenumbers in the range of about 12,800–30 cm⁻¹. Depending on the wavelength of this radiation, fingerprint spectra of molecular structures can be obtained. This data provides information on the structure of the molecule and, in particular, the nature of the functional groups present in the sample. The relationship of the different forms of IR analytical techniques to the rest of the electromagnetic (EM) spectrum are shown in Figure 6.1.

6.2.1 NEAR-INFRARED SPECTROSCOPY

Near-infrared (NIR) spectroscopy is a high-energy vibrational technique, covering the transition from the visible spectral range to the MIR region. The NIR region was first discovered by William Herschel in the 1800s, but it was not until the

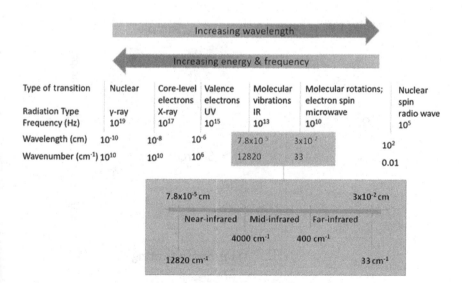

FIGURE 6.1 Electromagnetic spectrum and infrared region.

1950s that the spectroscopy technique was first used for analytical applications (Pasquini 2018). This rapid and non-destructive technique can be used to analyse pharmaceuticals with little or no sample preparation (Roggo et al., 2007; Lohumi et al., 2015; Si et al., 2021). The non-destructive nature of the technique allows samples to be subsequently analysed by other analytical techniques, or if acquisition is through the packaging, there is a possibility to return the samples into circulation (Roggo et al., 2007; Rodionova et al., 2018). Spectral data can be obtained through packaging materials, including glass and blister packaging (Krakowska et al., 2016).

For NIR spectroscopic analysis, samples are illuminated with a broad spectrum of NIR radiation that can be absorbed, transmitted, reflected, or scattered by the sample (Figure 6.2). The use of a prism or grating separates the frequencies emitted from the source, while a detector simultaneously measures the amount of energy that passes through (Lohumi et al., 2015). NIR spectroscopy uses overtone and recombination bands in the range of 750–2500 nm (12,820–4000 cm^{-1}) to determine the structure of the sample (Krakowska et al., 2016; Pasquini 2018). NIR spectra comprise of overtones and combination bands that are mainly attributed to hydrogen vibrations (CH, NH, OH) (Reich 2005; Lohumi et al., 2015). Absorption of NIR radiation in the matter is usually not uniform and depends on the molecular structure and number of bonds in the molecule (see Figure 6.2).

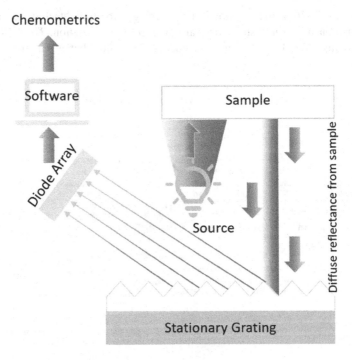

FIGURE 6.2   Schematic of a near-infrared spectrometer.

The properties of a sample can be investigated using different spectral modes, these assorted modes offer a variety of information about the sample. The interaction can be reflectance (specular and diffuse), transmittance, interactance, and transreflectance (Pasquini 2018). The spectral mode employed depends on the physical properties of the sample. Reflectance or interactance is generally used for solid samples, whereas transmission is used for liquids. Thin or clear samples are analysed using transreflectance (Lohumi et al., 2015). The most common mode applied to solid samples is diffuse reflectance (Pasquini 2018); this occurs when the measured light is reflected from a rough surface. The data generated contains information about the chemical and physical properties of the sample (Lohumi et al., 2015). The spectral data produced provides molecular information not only of the API(s), but also of other excipients present from a single measurement (Laasonen et al., 2004; Baer et al., 2007). While benchtop NIR spectrometers have been widely studied for assessing the quality of pharmaceuticals, portable and handheld versions are being investigated for this application. The ability of portable NIR spectrometers allows in the in-field real time screening to be performed.

A disadvantage of NIR spectroscopy is that only broad bands are observed and the spectra can be difficult to interpret due to overlapping signals (Roggo et al., 2007). The observed absorbance bands are the result of overtones from different functional groups within the sample analysed. Therefore, molecules with a similar chemical structure may be difficult to distinguish. With NIR spectroscopy, mixture analysis can present spectra with superimposed absorptions and compounds and may only be recognised if they have a unique functional group (Baer et al., 2007). Furthermore, careful calibration and acquisition of standard spectra of APIs and excipients is required in order to ensure accurate quantitative identification (Pasquini 2018). NIR spectroscopy may also not be sensitive enough to detect slight manufacturing differences between pharmaceutical samples (Rebiere et al., 2021). It has been implied that storage conditions of pharmaceuticals should be considered when using NIR spectroscopy for their analysis since NIR spectra are sensitive to samples that absorb water from the atmosphere (de Peinder et al., 2008; Moffat et al., 2010). The lack of specificity of the data in an NIR spectrum combined with the complexity of the source of the signal means that chemometrics approaches are always required to determine the sample composition and to support the authentication of a pharmaceutical product (Roggo et al., 2007).

6.2.2 MID-INFRARED SPECTROSCOPY

The electromagnetic energy of molecular vibration is defined as the infrared region, or MIR, in the range of from 4000–400 cm^{-1}. MIR spectroscopy has benefitted greatly from the development of microcomputer and spectroscopic approaches based on Fourier transform, thus giving rise to Fourier transform infrared (FTIR).

FTIR spectroscopy is a prominent vibrational technique and is a sophisticated tool in the spectral analysis of organic compounds (Cheng et al., 2010; Jamwal et al., 2021). FTIR spectroscopy is used to obtain structural information about a compound, has many advantages as an analytical tool, and is a reliable analytical technique for qualitative and quantitative analysis (Jamwal et al., 2021). This is because all APIs have an IR spectrum; the fingerprint region of the spectrum is unique and can be used for qualitative identification of medicines (Cheng et al., 2010; Lohumi et al., 2015). The fingerprint region is 2000–400 cm^{-1}; the vibrational frequencies and individual molecular motions observed in this range are unique to a particular molecule. FTIR spectra are information rich, provide a greater amount of chemical information, and are easier to interpret than NIR spectra due to peak specificity. Therefore, the likelihood of misidentification using FTIR is less compared to NIR techniques (Roggo et al., 2007; Lohumi et al., 2015).

Skilled personnel are needed for sample preparation using transmission FTIR as this involves dispersing about 0.1%–2% w/w of the sample in a potassium bromide (KBr) matrix and pressing it into a disc before analysis. The strength of the resulting IR spectra is dependent on the amount and homogeneity of the powder sample dispersed in the matrix as well as the constant disc pathlength, which can be a major challenge to prepare. Therefore, ensuring the right amount of sample in the matrix is crucial in order to obtain reproducible and well-defined FTIR spectra. This sample preparation approach can be replaced by attenuated total reflectance (ATR) techniques. The increased sensitivity of FTIR spectrometers make ATR-FTIR spectroscopy a simple analytical technique (Cheng et al., 2010; Ortiz et al., 2013) for screening counterfeit or substandard pharmaceutical products.

6.2.2.1 Attenuated Total Reflectance Fourier Transform Infrared Spectroscopy

ATR-FTIR spectroscopy has the ability to measure a wide variety of solid and liquid samples without the need for complex and time-consuming sample preparation. ATR is achieved by IR light being totally internally reflected at the interface between the sample and the ATR crystal (Chan et al., 2003).

Figure 6.3 illustrates the principle of ATR-FTIR spectroscopy. The penetration depth depends on the wavelength, the refractive indices of the ATR crystal and the sample and the angle of the IR beam. This typically can be in the range of 0.5–5 μm (Roth et al., 2019). The light penetrating the sample is attenuated by interaction with characteristic molecular vibrations to produce a conventional IR spectrum (Chan et al., 2003). It is, therefore, necessary to ensure maximum surface contact between the sample and the ATR crystal. This necessity is a limiting factor of the ATR-FTIR technique.

ATR-FTIR spectroscopy is a powerful technique, which has successfully been used to analyse a wide range of samples. Unlike NIR spectroscopy, ATR-FTIR spectroscopy is unable to easily analyse pharmaceutical samples through their packaging and this is due to the low penetration depth of this IR instrumentation (Roth et al., 2019). This low penetration depth means that it does not draw the test sample the bulk of a sample, meaning ATR-FTIR is classed as a surface technique (Bugay 2001; Planinšek et al., 2006). This may lead to misleading analysis results of whole tablets and capsules; therefore, tablets must be ground into fine powders

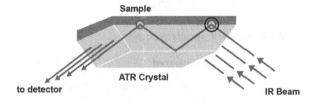

FIGURE 6.3 Principle of ATR-FTIR spectroscopy.

and capsule contents emptied and then analysed in order to generate reproducible spectra (Roth et al., 2019). Some studies of the ATR-FTIR analyses of tablet dosage forms have shown that while the stated dosage is correct, the distribution of the API throughout the tablet is not uniform (Lawson et al., 2014; Armitage 2018). Homogenisation of the samples for quantitative analysis is therefore essential. One factor that should be taken into consideration is that while powders are more suited for ATR-FTIR analysis, signal variations can be introduced into the spectrum, as the amount of pressure applied across the ATR crystal surface can vary (Ogwu 2018; Roth et al., 2019).

Powder-mixing efficiency is an important factor in any quantitative study using ATR-FTIR spectroscopy. Salari and Young (1998) found that a perfect mixture was 'virtually unattainable with powders', though it was possible to achieve a mixture with a maximum degree of randomness. They also cited concerns that interfacial properties of the original material could be replaced when mixed with a lubricant that was able to coat the powder surface. Scanning electron microscopy (SEM) was used to observe the organisation of the mixture and to see if a perfect mixture could be attained. It was also found that water content influenced the agglomeration of particles (Salari and Young 1998).

6.2.3 COMPARISON OF NIR AND ATR-FTIR SPECTROSCOPY

When comparing the two IR analytical approaches for the effective and rapid surveillance and monitoring of counterfeit and substandard medicines, NIR spectroscopy offers a non-destructive technique that can be used to analyse pharmaceuticals and/or packaging directly without sample preparation, whereas ATR-FTIR analyses require powder mixtures from solid pharmaceutical dosage forms. The two IR techniques are complementary, can be used in a tandem and are available as both benchtop and portable instruments. Benchtop systems have been widely implemented for analysing pharmaceuticals; however, these systems can be cost prohibitive and require extensive training. Handheld and portable spectrometers are often viewed as an affordable rapid screening tool in low-resource countries (Eady et al., 2021). Table 6.1 compares the key characteristics of the two IR techniques. Significant differences can be observed in the selectivity of the spectra of NIR and ATR-FTIR systems; the NIR spectra contain broad overlapping

TABLE 6.1
Comparison of NIR and MIR spectroscopy

Characteristic	Near-infrared	Mid-infrared
Principle	Overtone and combination absorptions	Fundamental absorptions
Sampling Mode	Diffuse reflectance	ATR
Sample Configuration	Stand-off detection	Intimate contact with sample
Sample Pathlength	~1–5 mm	~0.5–5 μm
Minimum Sample Size	~50 mg	~1 mg
Excitation Source	Broadband IR	Broadband IR
Chemical Sensitivity	Polar bonds	Polar bonds
Selectivity	Low (Chemometrics is required)	High
Speed of Analysis	Rapid	Rapid
Type of System	Benchtop Portable Handheld	Benchtop Portable
Approximate Purchase Cost	>US$20,000	>US$25,000

peaks, whereas ATR-FTIR spectra offer greater peak specificity. Figure 6.4 shows the NIR and ATR-FTIR spectra for an atenolol tablet.

6.3 CHEMOMETRIC APPROACHES USED IN COMBINATION WITH IR SPECTROSCOPY

Chemometrics is an interdisciplinary discipline that uses mathematical and statistical methods for the extraction and data treatment of chemical analyses (Jamwal et al., 2021). Wold and Kowalski are credited with first to use the term 'chemometrics' in 1972 (Bovens et al., 2019). The field of chemometrics rapidly developed with the advancement of computer and software technology in the 1980s (Bovens et al., 2019), which coincided with the development of NIR spectroscopy as an analytical tool (Pasquini 2018). Since NIR spectra are seen as a vast source of analytical information, but are very complex, various algorithms have been developed to overcome barriers imposed by the lack of selectivity in the NIR spectrum (Pasquini 2018). Over the past decade, the combination of ATR-FTIR spectroscopy and chemometrics has also been used to aid discrimination and classification of counterfeit and substandard medicines.

The workflow for general chemometric data consists of several steps of equal importance. These include pre-treatment, classification methods (unsupervised and supervised), and regression methods (non-linear and linear) (Custers et al., 2015; Krakowska et al., 2016; Jamwal et al., 2021). Classification methods, in general,

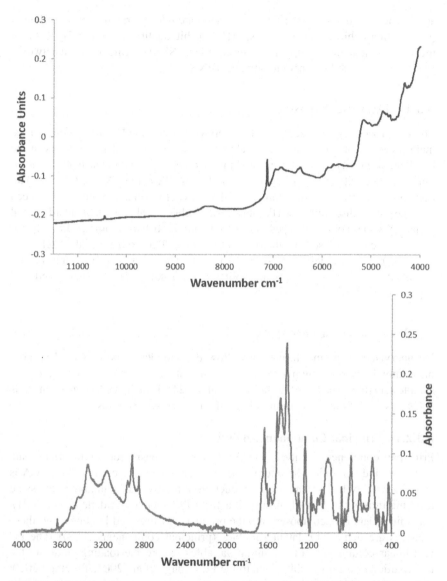

FIGURE 6.4 (A) NIR spectrum of an atenolol tablet and (B) ATR-FTIR spectrum of an atenolol tablet.

are used to predict qualitative properties and regression methods are the main approach for quantitative analysis (Biancolillo and Marini, 2018).

The following section discusses the more common methods used for chemometric analysis. Some NIR and ATR-FTIR applications for the rapid detection of counterfeit medicines have employed less common chemometric approaches, as detailed in Table 6.2. These include; correlation in wavelength (CWS), hierarchical cluster analysis (HCA), hit quality index (HQI), K fuzzy means, silhouette index, support vector machines (SVM), artificial neural networks (ANN) and probabilistic neural networks (PNN).

6.3.1 Data Pre-Processing

The pre-processing of spectra is a key step to extract reliable information and the main objective of this process is to eliminate the influence of noise, shifts in the baseline, variations in the detector and influence of solid particle size as well as to improve selectivity (Pasquini 2018; Tie et al., 2020; Xu et al., 2020). Pre-treatment transforms the raw data, therefore, the risk of loss of information must be balanced with the gain in discrimination (Pasquini 2018; Rebiere et al., 2021). Standard normal variate (SNV) is commonly used to remove variations between samples and can also be used to remove noise from the spectra (Xu et al., 2020; Rebiere et al., 2021).

Further data modeling is often required. The next step is often to use classification models that can be split into two categories; unsupervised and supervised methods.

6.3.2 Unsupervised Methods

Unsupervised chemometric methods allow the samples to be classified without prior knowledge and attempt to identify similarities and differences between the sample set (Roggo et al., 2007; Jamwal et al., 2021). Principal component analysis (PCA), is one of the most commonly used unsupervised methods.

6.3.2.1 Principal Component Analysis

Principal component analysis (PCA) forms the basis for multivariate data treatment and simplifies the interpretation of such data (Tie et al., 2020). PCA is an effective tool for dimensionality reduction in multivariate data, but at the same time minimises the loss of information from the data set (Aidene et al., 2021). The initial data is decomposed into the product of scores and loadings, and these matrices are defined by latent variables (principal components (PCs)). The PC is a linear combination of the initial variables, where the loadings indicate the contribution of each variable given to the PC (Roggo et al., 2007; Tie et al., 2020; Aidene et al., 2021). The higher the loading value, the more significant the variable. The first PC is designed to explain the maximum variance in the data, the second PC is orthogonal to the first and relates to the largest residual variation around PC1 (Tie et al., 2020; Rebiere et al., 2021). Scores are defined as the values of the projections of the samples on the new coordinates. The scores obtained from the PCA plot allow information about the similarities and differences among samples to be visualised (Custers et al., 2015; Tie et al., 2020; Aidene et al., 2021).

The PCA plot also allows the clustering pattern within the data to be visualised, as samples with similar properties are expected to cluster near to each other

(Jamwal et al., 2021). If the PCA plot shows discriminating tendencies but effective discriminations cannot be determined, then supervised analyses may enhance this (Storme-Paris et al., 2010).

6.3.3 SUPERVISED METHODS

Supervised chemometric methods differ to unsupervised methods, in that prior knowledge of the particular class or category of the sample is known, with targeted focus on the similarities within the different classes (Storme-Paris et al., 2010). A classification model is generated using a training set of samples, which can be used to evaluate and predict classes of an unknown sample from an independent validation set (Jamwal et al., 2021). Supervised methods focus on discrimination methods, such as linear discriminant analysis (LDA), and those that correlate similarities within a class, such as soft independent modeling of class analogies (SIMCA).

6.3.3.1 Linear Discriminant Analysis

Linear discriminant analysis (LDA) is a linear and parametric chemometric method with discriminating characteristics that use Euclidean distance to maximise separation between classes and minimise within class-variance. LDA principally looks to identify optimal boundaries between classes (Roggo et al., 2007; Jamwal et al., 2021). LDA is similar to PCA as both methods look to reduce features in the data using latent axes to explain variance in the data. The two chemometric methods differ in that PCA selects a direction to retain maximum structure in a lower dimension, whereas LDA selects directions that allows for maximum separation among the different classes (Roggo et al., 2007). The combination of these two chemometric methods allows the efficiency of the classification to be improved (Xu et al., 2020). LDA is limited by the number of variables in the data, which must be smaller than the number of observations (Jamwal et al., 2021).

LDA can be further modified if the distribution of variables does not have a similar shape, this is known as quadratic discriminant analysis (QDA). QDA is a non-linear method and assumes that the groups have normal multivariate distributions but with different variance (Miller and Miller 2018).

6.3.3.2 Soft Independent Modelling of Class Analogies

Soft independent modelling of class analogies (SIMCA) is a parametric method and the main class modelling technique that emphasises the similarities within a class rather than discrimination between classes. It is based upon PCA and uses it's modelling properties (Tie et al., 2020; Xu et al., 2020). Each class is independently modelled and focuses on the target class. For each class a PCA model is created, determining the optimum number of PCs required from each class to generate a cross-validation procedure. The capability of a model to select the number of PCs is estimated by utilising leave one out cross validation (LOOCV) (Jamwal et al., 2021). Two critical values need to be taken into consideration when constructing classification rules: (1) Euclidean distance towards the SIMCA model; and (2) Mahalanobis distance determined in the space of scores. These two significant values determine a restricted space around the objects (Custers et al., 2015).

The location of an unknown sample is determined by the scores and loadings of the PCA models, and can be classified into a particular class if it is located within a space defined by Euclidean and Mahalanobis distances (Custers et al., 2015; Tie et al., 2020). The robustness of the calibration set for each class and the differences between classes define the performance of an SIMCA model (Jamwal et al., 2021). SIMCA is a soft classification method – a disadvantage of this, is that an unknown sample could be assigned to more than one class (Xu et al., 2020).

6.3.3.3   Data Driven-Soft Independent Modelling of Class Analogies

Data driven-soft independent modelling of class analogies (DD-SIMCA) is a modern version of SIMCA (Si et al., 2021). DD-SIMCA is used to develop a threshold, which eliminates the target class from all other samples. This chemometric tool offers a threshold model that can calculate type I α and type II β errors in the data. Training sets are calculated using the number of samples and variables. The number of PCs determines the complexity of the model, which, in turn, influences the quality of classification (Rodionova et al., 2018). DD-SIMCA calculates the Euclidean (orthogonal) and Mahalanobis (score) distances for each object from the training set and calculates an acceptance threshold defined by the given value of the type I α error. If alternative classes are available, the method allows for the type II β error to be calculated with respect to individual classes. The end results enable a sample to be characterised by the complexity of the PC model and the threshold for a given α value. A significant characteristic of the model quality is calculated between the α value and sensitivity. To create the model both training and test sets are required, and ideally more than one set should be used in the modeling (Rodionova et al., 2018).

6.3.3.4   Partial Least Squares-Discriminant Analysis

Partial least squares-discriminant analysis (PLS-DA) is a supervised projection technique that is parametric and a linear method. Similar to PCA, PLS identifies latent variables known as PLS factors in featured spaces that have maximum covariance within the predicted variables. PLS-DA is used to differentiate between a group of objects when the response vectors are categorical variables. The variables are constructed to allow covariation between the original variables and response variables to be represented (Custers et al., 2015; Tie et al., 2020).

6.3.3.5   k-Nearest Neighbour

k-Nearest neighbour (k-NN) is a nonparametric and discriminant chemometric method (Roggo et al., 2007; Custers et al., 2015) that uses nearest neighbours to classify objects (Tie et al., 2020). A construction model is constructed by calculating the Euclidean distance (correlation coefficient) between an unknown object and each object of the training set (Tie et al., 2020). The k-nearest objects with the highest correlation coefficient are selected and a majority rule is applied. The unknown object is attributed to the class, which the largest number of k-nearest neighbouring objects of the training set belongs to (Custers et al., 2015; Tie et al., 2020).

Classification models are optimised to determine the number of nearest neighbours (k) to be included. This can mean that several models are built using different values for k; these models are further validated by a cross-validation procedure, allowing the best model to be selected (Custers et al., 2015). k-NN is best applied to solve binary problems, as it can distinguish between compliant and noncompliant data sets (Tie et al., 2020).

6.3.3.6 Classification and Regression Tree

Classification and regression tree (CART) is used to solve classification and regression problems, depending on the response vector defined. A classification tree is generated if the vector is categorical; if it is continuous, then a regression tree is created (Custers et al., 2015). CART analysis can be performed in three sequential steps. Step one creates a maximum tree, where all objects are included and each branch is split using a binary split procedure. This is where, based upon the variable, a mother group is split into two daughter groups. This procedure is repeated until a maximum tree is generated; the tree often shows overfitting of the data. Step two 'prunes' the maximum tree by cutting the terminal branches, creating several smaller and less complex trees. The optimal tree is created in step three and is selected upon the cross-validation of the predictive error. CART is advantageous, as it offers a clear visualisation of the generated model, resulting in a manifest set of decision rules (Custers et al., 2015).

6.3.4 REGRESSION METHODS

Regression methods are applied to relate the data to the quantifiable properties of the sample. These methods are developed by dividing the samples into two subsets – training or validation (Jamwal et al., 2021). The training set is built on a mathematical model by various regression methods, which is validated through the validation set. These sets can be used to calculate the standard error of calibration (SEC) of the model. Validation of the model is important, as it tests the ability of the model to predict new samples (Jamwal et al., 2021). Regression methods include principal component regression (PCR), partial least squares regression (PLS-R) and CART.

6.3.4.1 Principal Component Regression

PCR is a widely used regression technique and involves spectral data being decomposed by using PCA, multi-linear regression (MLR) analysis is performed on the PC score as predicted values. The advantage for using PCA means that any spectral co-linearity is supressed by omitting vectors that have a small magnitude.

However, a disadvantage is that the computed PCs are not guaranteed to correlate to the observed characteristic (Roggo et al., 2007).

6.3.4.2 Partial Least Squares Regression

PLS-R is the most widely applied multivariate method and is closely related to MLR and PCR analysis (Jamwal et al., 2021). PLS-R uses the least squares algorithm to compute regression models, where covariance is maximised by correlating the variance and data together. The aim is to establish a linear link between the spectral data and the reference value matrices (Roggo et al., 2007). PLS-R only includes the variables that are important for the ability to explain variations in the chemical data (Jamwal et al., 2021).

Both PCR and PLS-R provide similar outcomes and perform data decomposition into loadings and scores. The two regression methods differ in that PCR only decomposes the spectral data, whereas PLS-R applies to both spectral and concentration data. PLS-R also differs in that it uses fewer latent variables to build a model than PCR, although this does not affect the overall predictive ability (Jamwal et al., 2021).

6.4 APPLICATIONS OF IR TECHNIQUES FOR THE DETECTION OF COUNTERFEIT AND SUBSTANDARD MEDICINES

Near-infrared (NIR) and MIR spectroscopy analytical technologies offer quick, robust, low-cost, easy to use, and portable methods to discriminate between genuine, substandard, and counterfeit pharmaceuticals. Table 6.2 contains a summary of literature references where these analytical technologies have been applied in qualitative and quantitative analyses of pharmaceutical dosage forms and/or packaging for assessing medicine quality. As can be seen in Table 6.2, NIR and ATR-FTIR devices have been extensively investigated over the past two decades in pharmaceutical forensic analysis applications. Almost all therapeutic areas of medicines have been investigated using these IR devices, ranging from antibiotics (Storme-Paris et al., 2010; Khan et al., 2016; Alotaibi et al., 2018; Chen et al., 2018; Chen et al., 2020; Assi et al., 2021; Mittal et al., 2021), antimalarials (Ricci 2007; Dowell et al., 2008; Marson et al., 2016; Ogwu 2018; Ciza et al., 2019; Wang et al., 2019), antivirals (Lopes and Wolff, 2009), cardiovascular therapy medicines (de Peinder et al., 2008; Armitage 2018; Hattori et al., 2018; Assi et al., 2019), anti-diabetic medicines (Farouk et al., 2011; Armitage 2018), antidepressants (Storme-Paris et al., 2010), antiviral agents (Lopes and Wolff, 2009; Wang et al., 2019), antituberculosis medicines (Wang et al., 2019; Rodionova et al., 2021), antifungal medicines (Rodionova et al., 2019) and antihistamines (Rodionova et al., 2018) to cheap, over-the-counter analgesics including paracetamol (Said et al., 2011; Lawson et al., 2018; Ogwu 2018; Ciza et al., 2019). Lifestyle medicines including Cialis and Viagra have also been extensively investigated using these IR devices (O'Neil et al., 2008; Venhuis et al., 2010; Sacre et al., 2010; Assi et al., 2011; Deconinck et al., 2012; Anzanello et al., 2013; Ortiz et al., 2013; Anzanello et al.,

2014a; 2014b; Custers et al., 2015; Kahmann et al., 2018; Said et al., 2019; Brito et al., 2020).

6.4.1 NEAR-INFRARED SPECTROSCOPY APPLICATIONS

As is evident from Table 6.2, NIR spectroscopy has been extensively investigated for determining pharmaceutical sample authenticity via direct analysis of tablet and capsule (solid dosage) formulations, although most investigations were qualitative. One of the earliest NIR spectroscopy applications for detecting counterfeit medicines employed PCA chemometric algorithms for the screening of Viagra tablets. This analysis revealed counterfeit Viagra samples with a high level of accuracy and revealed the presence of other substances in the tablet as well as differences in sample homogeneity and composition (Vredenbregt et al., 2006).

Several applications of NIR spectroscopy for the rapid detection of counterfeit medicines have employed PCA followed by SIMCA as chemometric tools for the qualitative analysis/authenticity of pharmaceutical dosage forms. One of the earliest incisive investigations employing these multivariate modelling and classification algorithms showed unequivocally the potential of NIR spectroscopy for rapid, in-field and non-destructive identification of counterfeit pharmaceuticals from a variety of solid and liquid dosage forms. This study investigated the effect of tablet-sample face and position in relation to the probe beam together with effect of humidity and demonstrated NIR spectroscopy to be capable of identifying subtle alterations in drug composition (Scafi and Pasquini 2001). The discriminating ability of NIR spectroscopy was also proven with the unsupervised PCA chemometric method for counterfeits and imitations screening of fluoxetine and ciprofloxacin tablets, whereas the NIR spectroscopy discrimination ability between close formulations of generics required the use of the supervised SIMCA chemometric technique (Storme-Paris et al., 2010). Said and co-workers created a NIR spectroscopy spectral database of paracetamol and used PCA/SIMCA multivariate modelling and classification algorithms to quickly classify Malaysian paracetamol tablet samples according to their type of dose, dosage forms, and product variability. They reported some ambiguity in the results that could be resolved by further chemometric analysis and/or other analytical techniques (Said et al., 2011). Atorvastatin tablet samples were also investigated by NIR spectroscopy along with PCA and SIMCA chemometric approaches (Hattori et al., 2018). This study showed it was possible to differentiate between branded and generic versions of the atorvastatin tablets using this qualitative NIR spectroscopy method and that classification of the tablet samples was dependent on the excipient combinations. Near-infrared spectroscopy offers the advantage of providing a picture of the whole dosage form instead of just the API content. Combining NIR spectroscopy with PCA and SIMCA for the investigation of tablets and capsule contents of other antibiotics (amoxicillin and clavulonic acid, azithromycin, ciprofloxacin, doxycycline, and oflaxacin) also confirmed that this portable screening tool could distinguish between branded and generic medicines and could classify medicines according to their manufacturing

TABLE 6.2
Application of infrared spectroscopy methods for the detection of counterfeit and substandard medicines

IR spectroscopic technique	Sample type	Medicine(s) / Lifestyle drug	Qualitative / Quantitative analysis	Chemometric method(s)	Additional analytical technique	Reference
ATR-FTIR	Blister packaging	-	Qualitative	-	DSC	Salim et al., 2021
ATR-FTIR	Powder	Amoxicillin, Azithromycin, Cefaclor, Cefadroxil, Cefdinir, Cefpodoxime, Cefprozil, Cefuroxime, Cephalexin, Ciprofloxacin, Clarithromycin, Clindamycin, Levofloxacin, Linezolid, Metronidazole, Minocycline, Moxifloxacin, Nalidixic acid, Nitrofurantoin, Norfloxacin, Ofloxacin, Rifaximin, Roxithromycin, Sparfloxacin, Sulfasalazine, Tetracycline, Tinidazole	Qualitative	PLS-DA PCA	-	Mittal et al., 2021

Technique	Sample	Active ingredient	Type	Chemometrics	Other techniques	Reference
NIR	Tablet, Capsule contents	Amoxicillin and Clavulonic acid, Azithromycin, Ciprofloxacin, Doxycycline, Ofloxacin	Qualitative	PCA, SIMCA	-	Assi et al., 2021
NIR	Capsule shell	Terizidone	Qualitative	PCA, DD-SIMCA	-	Rodionova et al., 2021
NIR	Tablet	Azithromycin, Roxithromycin, Levofloxacin, hydrochloride	Qualitative	PCA, DD-SIMCA	-	Chen et al., 2020
NIR	Tablet	Modafinil	Quantitative	PLS-R	Raman	Assi et al., 2020
ATR-FTIR	Powder	Cialis®, Viagra®	Qualitative	k-NN, CWS	-	Brito et al., 2020
ATR-FTIR, NIR	Powder, Tablet	Atenolol, Captopril, Clopidogrel, Enalapril, Losartan, Olmesartan, Propranolol, Valsartan	Qualitative	PCA	-	Assi et al., 2019
ATR-FTIR	Packaging - vials		Qualitative		Optical Microscopy, Raman, XRF, SEM	Dégardin et al., 2019
NIR	Tablet	Levofloxacin	Qualitative	PCA, DD-SIMCA	-	Chen et al., 2019
NIR	Capsule shell, Blister packaging	Fluconazole	Qualitative, Quantitative	DD-SIMCA, PLS-R	-	Rodionova et al., 2019

(continued)

TABLE 6.2 (Continued)
Application of infrared spectroscopy methods for the detection of counterfeit and substandard medicines

IR spectroscopic technique	Sample type	Medicine(s) / Lifestyle drug	Qualitative / Quantitative analysis	Chemometric method(s)	Additional analytical technique	Reference
NIR	Blister packaging	Artemeter-lumefantrine, Paracetamol, Ibuprofen	Qualitative	HCA DD-SIMCA HQI	Raman	Ciza et al., 2019
NIR	Tablet	Viagra®	Qualitative	PCA	Mass spectrometry NMR	Said et al., 2019
NIR Mid-infrared (MIR)	Powder	Artemether, Efavirenz, Isoniazid	Quantitative	Multivariate PLS-R	-	Wang et al., 2019
ATR-FTIR	Powder	Chloroquine, Paracetamol	Qualitative Quantitative	Multivariate PLS-R	UV-Vis	Ogwu 2018
ATR-FTIR	Packaging – boxes, vials and leaflets	-	Qualitative	-	Raman XRF SEM microCT	Dégardin et al., 2018
NIR	Tablet	Atorvastatin	Qualitative	PCA SIMCA	-	Hattori et al., 2018
NIR	Blister packaging	Loratadine	Qualitative	DD-SIMCA	VisCam (visual analysis)	Rodionova et al., 2018
NIR	Tablet	Metronidazole	Qualitative	PCA PLS-DA	-	Chen et al., 2018
ATR-FTIR	Powder	Paracetamol	Qualitative Quantitative	Multivariate PLS-R	UV-Vis	Lawson et al., 2018

Technique	Sample	Drug	Qualitative/Quantitative	Multivariate method	Reference method	Reference
ATR-FTIR	Powder	Amoxicillin	Qualitative Quantitative	Multivariate PLS-R	HPLC-UV	Alotaibi et al., 2018
ATR-FTIR	Powder	Atenolol, Metformin	Qualitative Quantitative	Multivariate PLS-R	UV-Vis	Armitage 2018
ATR-FTIR	Powder	Cialis®, Viagra®	Qualitative	k-NN LDA SVM	–	Kahmann et al., 2018
ATR-FTIR	Single droplet of ampoule content	Durateston	Qualitative	PCA PLS-DA	GC-MS	Neves et al., 2017
NIR	Tablet	–	Qualitative	PCA k-NN SVM DA	–	Dégardin et al., 2016
IR NIR	Blister packaging, Tablet	Cialis®, Viagra®	Qualitative	PCA CART SIMCA PLS-DA	–	Custers et al., 2016
NIR	Blister packaging	–	Qualitative	DD-SIMCA	–	Zontov et al., 2016
ATR-FTIR	Powder	Artesunate, Mefloquine	Qualitative Quantitative	Multivariate PLS-R	LC-MS/MS	Marson et al., 2016
NIR	Capsule	Amoxicillin	Qualitative Quantitative	PLS-DA	HPLC	Khan et al., 2016
ATR-FTIR	Powder	Cialis®, Viagra®	Qualitative	PCA k-NN CART SIMCA	–	Custers et al., 2015

(continued)

TABLE 6.2 (Continued)
Application of infrared spectroscopy methods for the detection of counterfeit and substandard medicines

IR spectroscopic technique	Sample type	Medicine(s) / Lifestyle drug	Qualitative / Quantitative analysis	Chemometric method(s)	Additional analytical technique	Reference
NIR	Liquid	Quinine	Qualitative Quantitative	PCA Multivariate PLS-R	Raman HPLC	Mbinze et al., 2015
ATR-FTIR	Powder	Cialis®, Viagra®	Qualitative	PCA k-means Fuzzy C-means Silhouette Index	–	Anzanello et al., 2014a
NIR	Tablet	Anisodamine	Qualitative	PCA PLS-DA	Raman	Li et al., 2014
ATR-FTIR	Powder	Cialis®, Viagra®	Qualitative	PCA k-NN SVM	XRF ESI-MS UPLC-MS	Anzanello et al., 2014b
ATR-FTIR	Powder	Cialis®, Viagra®	Qualitative	PCA SM	–	Ortiz et al., 2013
ATR-FTIR	Powder	Cialis®, Viagra®	Qualitative	PCA k-NN	–	Anzanello et al., 2013
ATR-FTIR	Powder	Diethylpropione, Fenproporex, Sibutramine	Qualitative	PCA	GC-MS UV-Vis	Mariotti et al., 2013

Technique	Sample	Drug	Type	Chemometrics	Other	Reference
FTIR, NIR	-	Cialis®, Viagra®	Qualitative	PCA, CART, k-NN	Raman	Deconinck et al., 2012
FTIR, NIR	Capsule contents – dried extracts	Miltefosine	Qualitative	PCA	LC-MS/MS	Dorlo et al., 2012
FTIR	KBr disc	Metformin, Pioglitazone, Repaglinide, Rosiglitazone	Qualitative, Quantitative	-	-	Farouk et al., 2011
FTIR, NIR	Powder Capsule	-	Qualitative	PCA, HCA, k-NN, PLS-DA, ANN, PNN	Raman, GC-MS	Been et al., 2011
NIR	Powder, Tablet	Atorlip, Cialis®, Clopivas, Dosan, Levitra, Plavix, Reductil, Rosiglitasone, Stocrin, Viagra®	Qualitative	CWS, PCA	-	Assi et al., 2011
NIR	Tablet	Paracetamol	Qualitative	PCA, SIMCA	-	Said et al., 2011
NIR	Capsule	Ciprofloxacin, Fluoxetine	Qualitative	PCA, SIMCA	-	Storme-Paris et al., 2010
FTIR, NIR	KBr disc, Tablet	Cialis®, Viagra®	Qualitative	PCA, PLS	Raman	Sacre et al., 2010
NIR	Blister packaging, Tablet	Cialis®	Qualitative	-	LC-DAD-MS, LC-UV-CD, NMR	Venhuis et al., 2010

(continued)

TABLE 6.2 (Continued)
Application of infrared spectroscopy methods for the detection of counterfeit and substandard medicines

IR spectroscopic technique	Sample type	Medicine(s) / Lifestyle drug	Qualitative / Quantitative analysis	Chemometric method(s)	Additional analytical technique	Reference
NIR	Tablet	Lamivudine	Qualitative Quantitative	PCA k-NN	-	Lopes and Wolff, 2009
NIR	Tablet	Artesunate	Qualitative Quantitative	Multivariate PLS-R	-	Dowell et al., 2008
NIR	Tablet	Cialis®, Levitra	Qualitative	PCA COM	-	O'Neil et al., 2008
NIR	Tablet	Atorvastatin Lovastatin	Qualitative	PCA PLS-DA	Raman	de Peinder et al., 2008
ATR-FTIR imaging	Blister packaging	Artesunate	Qualitative	-	Raman	Ricci 2007
NIR	Tablet	Viagra®	Qualitative	PCA	-	Vredenbregt et al., 2006
NIR	Tablet, Capsule, Solution, Paste	Acetylsalicylic acid, Potassium citrate Levonorgestrel and ethynylestradiol, Ampicillin, Enalapril maleate, Lactutone, Tioconazol and Tinidazol, Bromoprida and Dimeticona	Qualitative	PCA SIMCA	-	Scafi and Pasquini, 2001

sources (Assi et al., 2021). This investigation concluded that SIMCA provided more accurate classification over PCA for all antibiotics investigated except for ciprofloxacin products, which shared many overlapping excipients. Recent trends show that the newer multivariate chemometric tool, DD-SIMCA, is being applied as the discriminate approach for the NIR spectroscopy detection of counterfeit and substandard medicines (Zontov et al., 2016; Rodionova et al., 2018; 2019; 2021; Ciza et al., 2019). Studies by Chen and co-workers concluded that the combination of NIR spectroscopy, feature selection and DD-SIMCA class-modelling is feasible for the identification of expired tablets of the antimicrobial drug levofloxacin and for the rapid screening of azithromycin, roxithromysin and levofloxacin antibiotic tablets (Chen et al., 2019; 2020).

Further applications of NIR spectroscopy for the rapid screening of medicines have used PCA for exploratory data analysis followed by other supervised chemometric methods for discrimination and/or classification. In a study to screen medicines purchased from the internet and the global market the use of CWS and PCA methods simultaneously could identify potential counterfeits. The importance of authentic reference samples was highlighted in this comparative qualitative study (Assi et al., 2011). Combining the PCA and CWS chemometric methods was also explored to screen tablets of cardiovascular drugs (Assi et al., 2019). This study revealed that PCA on its own may not indicate authenticity where different manufacturing sources of products were involved; however, the CWS method was successful in authenticating a cardiovascular pharmaceutical product and raising alertness for potential counterfeit and/or substandard products. A study by Dégardin and co-workers employed NIR spectroscopy and several chemometric tools (PCA, k-NN, SVM, DA) for the rapid identification of 29 different pharmaceutical families of tablets. The available counterfeits, generics, and placebos were tested and rejected by the method, which confirmed the reliable use of the NIR spectroscopy method for the rapid analysis of suspect counterfeits of tablets (Dégardin et al., 2016). Classification of metronidazole tablet samples with respect to their brands was successfully achieved by NIR spectroscopy analysis coupled with PCA followed by PLS-DA, which indicated that the chemometrics model based on the reduced variable set outperformed the full-spectrum model (Chen et al., 2018).

Near-infrared spectroscopy has often been combined with Raman spectroscopy and chemometrics for counterfeit/substandard medicine detection. Classification based on PCA and supervised PLS-DA models (de Peinder et al., 2008; Sacre et al., 2010; Been et al., 2011; Li et al., 2014), CART, k-NN models (Been et al., 2011; Deconinck et al., 2012) were successful for both spectroscopic techniques. The rapid and cost-effective detection of counterfeit antimalarials is crucial in combating the threat of malaria especially in LIC and LMIC, where malaria is one of the main causes of child mortality. This pertinent issue was addressed in a study conducted in the Democratic Republic of Congo which compared the non-destructive NIR and Raman spectroscopy methods for the rapid identification of children's oral quinine drops purchased from local pharmaceutical markets. In this study, the spectroscopy data were validated with the conventional reference destructive, expensive and time-consuming HPLC assay for quinine dihydrochloride. The spectroscopy results indicated that it was possible to use PCA as a discriminating method to rapidly detect substandard samples and a successful study outcome was that the NIR spectroscopy qualitative model was to be implemented for routine analysis and replace the existing HPLC method in a national quality control laboratory of drugs (Mbinze et al., 2015). The feasibility of NIR and Raman spectroscopy to discriminate between genuine and counterfeits tablets of a cholesterol-lowering medicine, Lipitor was also investigated. Classification based on PLS-DA models was successful for both vibrational spectroscopic methods, irrespective of whether atorvastatin or lovastatin was used as the API. This study concluded that the discriminative power of the NIR spectroscopy model relies mainly on the spectral differences of the tablet matrix, which is due to the relatively large sample volume being probed with NIR radiation and the excipient's strong spectroscopic activity. It was further postulated that since Raman spectroscopy is primarily a surface technique, it poses a disadvantage for this type of analysis as the spectra for the tablet coating and core might differ and that the Raman spectra may change with the position of the laser for a non-homogenous sample (de Peinder et al., 2008). Sacre and co-workers aimed to compare and combine FTIR, NIR, and Raman spectroscopy for the detection of counterfeit Viagra and Cialis tablets. PLS analysis results revealed that for Viagra, the best results were provided by a combination of FTIR and NIR, while for Cialis the best results were provided by the combination of NIR and Raman spectroscopy (Sacre et al., 2010). Classification and regression tree algorithms were evaluated for modeling FTIR, NIR, and Raman spectroscopic data in order to discriminate between genuine and counterfeit tablet samples of Viagra and Cialis. k-NN was also applied to the same data sets. The results obtained with this tree-based method were easily interpretable and better than the more traditional k-NN discriminating method, although the CART models are limited by the nature of the data set and they should be adapted and updated each time new samples or classes are encountered (Deconinck et al., 2012). Rebiere and co-workers employed a two-step analytical approach to the rapid NIR and Raman spectroscopy non-destructive screening of anabolic tablets. Here the fast analysis of tablets using NIR spectroscopy to assess sample homogeneity based on their global composition, followed by Raman chemical imaging to obtain information on sample formulation was conducted. NIR spectroscopy assisted by PCA enabled fast discrimination of different profiles based on the excipient formulation and Raman imaging provided chemical images of the distribution of the active compound and excipients within tablets and facilitated identification of the active compounds (Rebiere et al., 2016). Another study evaluated the performances of various NIR and Raman spectroscopy handheld instruments in specific brand identification of tablet and capsule medicines through their primary packaging. The overall results showed good detection abilities for NIR instruments compared to Raman instruments, which are less sensitive to the physical state of the samples than the NIR systems (Ciza et al., 2019).

Quantification of the API in pharmaceutical dosage forms using NIR spectroscopy has been done mostly with multivariate PLS-R chemometric models (Mbinze et al., 2015; Rodionova et al., 2019; Wang et al., 2019; Assi et al., 2020). A study by Dowell and co-workers used NIR spectroscopy to accurately discriminate between counterfeit and genuine artesunate antimalarial tablets. Multivariate classification models indicated that this ability was based on the presence or absence of spectral signatures related to artesunate (Dowell et al., 2008). In a study by Lopes and Wolff, NIR chemical imaging (NIR-CI) was successfully used to detect counterfeit tablets of an antiviral drug, Heptodin, and to classify or source them so as to understand the possible origins. NIR-CI combined with multivariate analysis was well-suited to compare chemical and physical properties of the samples and for quantitative screening to determine the amount of the API lamivudine present in the tablet (Lopes and Wolff, 2009). Rodionova and co-workers emphasised that the correct API concentration in a dosage form does not necessarily guarantee that the medicine is a genuine one since a medicine can have a correct API concentration, but be produced by an illegal manufacturer. Therefore, they propose the NIR PLS-R calibration model they developed for the quantification of fluconazole concentration in capsules can be used as an additional, but not the sole technique of authentication (Rodionova et al., 2019). Wang and co-workers evaluated in both univariate and multivariate analyses, the antimalarial, antiviral and anti-tuberculosis API quantification performance of an NIR spectrometer, a handheld Raman spectroscopy device and a MIR spectrometer. From this exploratory study they established that certain NIR devices hold significant promise as cost-effective screening tools for counterfeit and substandard medicines (Wang et al., 2019).

More recently, the analysis of packaging by near NIR spectroscopy in combination with multivariate chemometric tools has shown promising results in the detection of counterfeit medicines. Rodionova and co-workers investigated NIR spectroscopy to assess capsule shell contents, blister packaging, and vials qualitatively and quantitatively and highlighted the importance of the chemometric tools for the detection of the poor-quality medicines (Rodionova et al., 2018; 2019; 2021). In earlier studies, MIR and NIR analyses of genuine and counterfeit Viagra and Cialis tablets in intact blister packaging was carried out and the application of SIMCA and PLS-DA chemometric tools showed perfect discrimination between genuine products and counterfeits (Custers et al., 2016).

6.4.2 MID-INFRARED SPECTROSCOPY APPLICATIONS

Most of the earlier investigations in the analysis of counterfeit medicines using MIR employed Fourier transform infrared (FTIR) spectroscopy (Sacre et al., 2010; Been et al., 2011; Farouk et al., 2011) as it eliminates the need for solvent extraction of the APIs/excipients. However, the requirement of skilled personnel to prepare KBr discs with a fixed pathlength and obtaining reproducible and well-defined FTIR spectra have been major challenges. Farouk and coworkers developed and validated FTIR quantitative assays, using a constant KBr disc pathlength, to quantify antidiabetic drugs in tablet dosage forms (Farouk et al., 2011).

Attenuated total reflectance Fourier transform infrared spectroscopy (ATR-FTIR) has revolutionized conventional FTIR by eliminating the main challenges in the analysis of pharmaceutical solid dosage forms, notably the time spent in sample preparation – sample extraction or KBr disc preparation and the lack of spectral reproducibility (Lawson et al., 2018). Given these advantages there has been a rise in number of applications of this technique for the rapid screening of medicine quality and for determining API content, as shown in Table 6.2. Prior to ATR-FTIR analysis, tablet and capsule dosage forms need to have been powdered to provide qualitative and/or quantitative data.

A significant number of ATR-FTIR applications have focused on the rapid screening of counterfeit Cialis and Viagra tablets (Anzanello et al., 2013; 2014a; 2014b; Ortiz et al., 2013; Custers et al., 2015; Kahmann et al., 2018; Brito et al., 2020). Principal component analysis followed by other supervised chemometric approaches such as k-NN, CART, Fuzzy C-mean, SM, SVM, and SIMCA have been applied for the qualitative analysis of Cialis and Viagra tablet samples and for the classification of tablets in these applications. From these studies, there was evidence that discrimination between authentic and counterfeit Cialis and Viagra tablets relied heavily on the information from excipient absorption bands (Anzanello et al., 2014a). To increase classification accuracy for Cialis and Viagra, it was suggested using data acquired from ultra-performance liquid chromatography-mass spectrometry (UPLC-MS) and physical profiling coupled with ATR-FTIR techniques (Anzanello et al., 2014b). Custers and co-workers showed that chemometric analysis of ATR-FTIR fingerprints is a valuable tool to discriminate genuine from counterfeit samples and to classify counterfeit medicines. Based on their results with Cialis and Viagra, SIMCA generated the best predictive models for classification. CART and k-NN were only useful for the discrimination based on the APIs present. Furthermore, they concluded that other chemometric methods would need to be explored for other classes of medicines (Custers et al., 2015). Since ATR-FTIR spectra typically result in a large number of wavenumbers, reducing the performance of classification chemometric methods aimed at discriminating between authentic and counterfeit medicines, Britto and co-workers proposed a novel wavenumber selection method (Britto et al., 2020).

Given the minimal sample preparation required for ATR-FTIR analysis, there has been increasing interest in applying this vibrational spectroscopy technique coupled with chemometric methods to rapidly screen tablet formulations qualitatively and quantitatively. Several studies have successfully developed, validated and applied multivariate PLS calibration models to determine paracetamol (Lawson et al., 2018; Ogwu 2018), antimalarials (Marson et al., 2016; Ogwu 2018), an antihypertensive (Armitage 2018), an antidiabetic drug (Armitage 2018), and antibiotics (Alotaibi et al., 2018) in tablets from the world market using the simple and rapid ATR-FTIR approach. The whole process of crushing, identifying, and quantifying the API in a tablet takes about five minutes per tablet sample after the ATR-FTIR method has been optimized. In these studies, the ATR-FTIR approach was compared to the time-consuming, costly and reference UV spectrophotometric, LC-UV or LC-MS/MS analyses, which require solvent extraction, dilution and filtration prior to analysis and generation of results. The ATR-FTIR method yielded comparable quantitative results. Identifying a suspect tablet sample is a priority in the rapid screening of medicines and this is the purpose for which ATR-FTIR spectroscopy was being assessed in these studies. The ATR-FTIR instrument employed in these studies is small and compact and its portability makes it valuable for in-field analysis, such as quality control by pharmaceutical companies and post-marketing surveillance by regulatory bodies. Furthermore, it is also relatively inexpensive and easy to use compared to the pharmacopoeia-approved techniques, so can potentially be used in LIC and LMIC where facilities and skilled personnel are not readily available.

Counterfeit and substandard antimicrobial drugs, including antibiotics, are a leading cause of antimicrobial resistance, particularly in LIC and LMIC. A study by Mittal and coworkers qualitatively assessed 57 antibiotic pharmaceutical products using ATR-FTIR spectroscopy combined with PCA and PLS-DA chemometric models (Mittal et al., 2021). The systematic approach presented together with the workflow proposed in this study could be used for fingerprinting of antibiotics based on functional groups for the screening of counterfeit antibiotic pharmaceutical products. The quick quantification of antibiotics in pharmaceutical dosage forms is also key in detecting substandard medicines and thus preventing antimicrobial resistance. The rapid quantification of the antibiotic amoxicillin using ATR-FTIR spectroscopy has been investigated using multivariate calibration models and showed that the field-portable ATR-FTIR instrument reliably identified substandard amoxicillin capsules and was found to yield quantitative results in good agreement with the established pharmacopeia HPLC-UV protocol (Alotaibi et al., 2018).

An ATR-FTIR method using PCA and PLS-DA chemometric methods was developed to detect counterfeiting of the anabolic steroid, Durateston, from a single liquid droplet from an ampoule. In this study, all Durateston ampoules were also analysed by GC-MS, which was the reference analytical method. The results showed that ATR-FTIR and PLS-DA is a suitable analytical tool for discriminating original samples from more elaborate counterfeits of Durateston and that it is a robust, cheaper and far less time-consuming alternative approach to the routine GC-MS analysis of suspect Durateston samples. ATR-FTIR can be easily implemented in all forensic laboratories to standardise and improve Durateston analysis (Neves et al., 2017).

Recent developments in the analysis of packaging materials by ATR-FTIR spectroscopy show promising results in the fight against counterfeit medicines. Salim and co-workers confirmed that the rapid analysis of polymeric blister packaging materials by ATR-FTIR spectroscopy and comparing differential scanning calorimetry (DSC) thermograms of the plastic in their packaging effectively discriminated counterfeit medicines (Salim et al., 2021). Dégardin and co-workers employed ATR-FTIR spectroscopy in combination with several analytical techniques to support authentication of pharmaceutical packaging (Dégardin et al., 2018; 2019). They concluded that ATR-FTIR spectroscopy combined with optical microscopy proved efficient to support the visual comparison for authentication of counterfeited flip off caps, aluminium crimping caps, stoppers, glass vials and vial labels. To detect links between counterfeits at each of these levels, the study also highlighted the importance of authenticating each part of the product.

6.5 CONCLUSION

Counterfeit and substandard medicines are widespread and a major impediment to improvements in public health globally. Developments over the past two decades in IR spectroscopic technologies and in chemometric methods have enabled the rapid discrimination of genuine and counterfeit/substandard medicines and can assist in the fight against this global public health challenge. NIR spectroscopy devices are significant non-invasive analytical tools for the rapid detection of poor-quality medicines, which can directly transmit through packaging components and possess good capacity for qualitative and quantitative analyses. ATR-FTIR devices are ideal analytical tools for the rapid fingerprinting of pharmaceuticals and have huge potential for semi-destructive and rapid quantitative analyses. These innovative portable devices employed to screen for poor-quality medicines can enhance rapid medicine surveillance and provide hope for the future. The optimised multivariate chemometric models might obviate the need for subsequent time-consuming, destructive and expensive chromatographic validation. The use of NIR spectroscopy for distinguishing branded and generic pharmaceutical products is fast and simple due to developments in chemometrics. To maximise the rapid detection and removal of counterfeit medicines from the pharmaceutical supply chain, and thus protect patients, an ideal portable IR screening device should have high sensitivity and be capable of detecting more than one API in a single pharmaceutical sample and be able to detect anomalies on primary packaging or products that are not apparent to the naked eye. This might be particularly relevant to protein or biotechnology-derived medicines, including vaccines, which normally require high-cost and time-consuming analytical techniques to exclude falsification. There is also the need for a uniform spectroscopy reference database which could be used on a wide array of spectroscopic devices, each with their own software.

Challenges remain in the worldwide accessibility of affordable, easy to operate, sensitive and portable IR spectroscopy detection devices for counterfeit and substandard medicines. In emerging economies or countries with limited resources, support for national medicines regulatory authorities is not always a priority, but investment in checking medicine quality on the ground is likely to have important beneficial effects on patient safety and public health and, thus, impact on the national clinical and economic burden. Making detection technologies more accessible and providing proper training to personnel in LIC and LMIC – where there is a large and pervasive spread of counterfeit and substandard medicines – are therefore essential. In LIC and LMIC, the lack of low-cost, mobile, easily operable, robust, accurate and validated testing methods to assess medicine quality in the field have restricted the ability of national medicines regulatory authorities, healthcare systems and enforcement authorities to detect and deal with the growing public health issue. Fast and accurate in-field routine analysis of pharmaceuticals and primary packaging is vital to develop an effective surveillance and monitoring system for the authentication of medicines in LIC and LMIC. The aim of the screening is to reduce the number of pharmaceutical samples that a national medicines quality control laboratory must test, which will reduce the burden on the laboratory and its limited resources. This is where IR spectroscopic tools – including ATR-FTIR and NIR spectroscopy devices – have a major role to play in the future in the fight against counterfeit and substandard medicines. They can be used for the initial in-field screening of the medicine and if further confirmatory testing is required then the samples can be sent to a well-equipped analytical laboratory. LIC and LMIC may not have access to such advanced laboratory facilities and skilled personnel to carry out the specialised confirmatory analyses. Therefore, collaboration between countries such that suspect pharmaceutical samples are sent to other well-equipped countries for confirmatory analysis may be a way forward in addressing the challenge of limited access to analytical facilities in some LIC and LMIC. The portable NIR and ATR-FTIR spectroscopy technologies have an added advantage since they do not require any expensive solvents and are therefore also sustainable and green analytical technologies to combat the global public healthcare challenge of counterfeit and substandard medicines.

6.6 IN MEMORIAM: DR GRAHAM LAWSON

This chapter is dedicated to Dr Graham Lawson (1946–2019). Dr Lawson was a well-respected analytical chemist internationally and inspired the co-authors of this chapter to conduct research on rapid analytical methods for the detection of counterfeit medicines. He was a role model, a great mentor, an inspirational colleague and a dear friend and is remembered with immense gratitude and respect.

6.7 ACKNOWLEDGMENTS

The authors would like to thank Dr Owen Wilkin (Bruker, UK) for the analysis of atenolol tablets by NIR spectroscopy.

REFERENCES

Aidene, Soraya, Maria Khaydukova, Galina Pashkova, Victor Chubarov, Sergey Savinov, Valentin Semenov, Dmitry Kirsanov, and Vitaly Panchuk. 2021. "Does chemometrics work for matrix effects correction in X-ray fluorescence analysis?" *Spectrochimica Acta Part B: Atomic Spectroscopy* 185, 106310. doi.org/10.1016/j.sab.2021.106310.

Alghannam, Abdulaziz F.A., Zoe Aslanpour, Sara Evans, and Fabrizio Schifano. 2014. "A systematic review of counterfeit and substandard medicines in field quality surveys." *Integrated Pharmacy Research and Practice* 4, no. 3: 71–86.

Almuzaini, Tariq, Imti Choonara, and Helen Sammons. 2013. "Substandard and counterfeit medicines: a systematic review of the literature." *BMJ Open* 3, no. 8: e002923. doi. org/10.1136/bmjopen-2013-002923.

Alotaibi, Norah, Sean Overton, Sharon Curtis, Jason W. Nickerson, Amir Attaran, Sheldon Gilmer, and Paul M. Mayer. 2018. "Toward point-of-care drug quality assurance in developing countries: comparison of liquid chromatography and infrared spectroscopy quantitation of a small-scale random sample of amoxicillin." *The American Journal of Tropical Medicine and Hygiene* 99, no. 2: 477–481. doi.org/10.4269/ajtmh.17-0779

Anzanello, Michel J., Rafael S. Ortiz, Renata P. Limberger and Paulo Mayorga. 2013. "A multivariate-based wavenumber selection method for classifying medicines into authentic or counterfeit classes." *Journal of Pharmaceutical and Biomedical Analysis* 83, 209–214. doi.org/10.1016/j.jpba.2013.05.004

Anzanello, Michel, Flavio S. Fogliatto, Rafael S. Ortiz, Renata Limberger, and Kristiane Mariotti. 2014a. "Selecting relevant Fourier transform infrared spectroscopy wavenumbers for clustering authentic and counterfeit drug samples." *Science & Justice* 54. doi.org/10.1016/j.scijus.2014.04.005

Anzanello, Michel J., Rafael S. Ortiz, Renata P. Limberger and Kristiane de Cássia Mariotti. 2014b. "A framework for selecting analytical techniques in profiling authentic and counterfeit Viagra and Cialis." *Forensic Science International* 235, 1–7. doi.org/ 10.1016/j.forsciint.2013.12.005

Armitage, Rachel. "Rapid screening methods to identify substandard and falsified medicines." PhD thesis, De Montfort University, Leicester, 2018.

Assi, Sulaf, Robert A. Watt, and Anthony C. Moffat. 2011. "Identification of counterfeit medicines from the internet and the world market using near-infrared spectroscopy." *Analytical Methods* 3, no. 10: 2213–2236. doi.org/10.1039/c1ay05227f

Assi, Sulaf, Ian Robertson, Thomas Coombs, Jacob McEachran, and Kieran Evans. 2019. "The use of portable near infrared spectroscopy for authenticating cardiovascular medicines." *Spectroscopy* 34, no. 5: 46–54.

Assi, Sulaf., Iftikhar Khan, Aaron Edwards, David Osselton, and Hisham Al-Obaidi. 2020. "On-spot quantification of modafinil in generic medicines purchased from the internet using handheld Fourier transform-infrared, near-infrared and Raman spectroscopy." *J Analytical Science and Technology* 11, no. 35. doi.org/10.1186/s40543-020-00229-3

Assi, Sulaf, Basel Arafat, Katherine Lawson-Wood, and Ian Robertson. 2021. "Authentication of antibiotics using portable near-infrared spectroscopy and multivariate data analysis." *Applied Spectroscopy* 75, no. 4: 434–444. doi.org/10.1177/0003702820958081

Bakker-'t Hart, Ingrid M.E., Dana Ohana, and Baastian J. Venhuis. 2021. "Current challenges in the detection and analysis of falsified medicines" *Journal of Pharmaceutical and Biomedical Analysis* 197, 113946. doi.org/10.1016/j.jpba.2021.113946.

Baer, Ines, Robert Gurny, and Pierre Margot. 2007. "NIR analysis of cellulose and lactose-application to ecstasy tablet analysis." *Forensic Science International* 167, no. 2–3: 234–241. doi.org/10.1016/j.forsciint.2006.06.056.

Beargie, Sarah M., Colleen R. Higgins, Daniel R. Evans, Sarah K. Laing, Daniel Erim, and Sachiko Ozawa. 2019. "The economic impact of substandard and falsified antimalarial medications in Nigeria." *PLoS ONE* 14, no. 8: e0217910. doi.org/10.1371/journal. pone.0217910

Béen, Frederic, Yves Roggo, Klara Dégardin, Pierre Esseiva and Pierre Margot. 2011. "Profiling of counterfeit medicines by vibrational spectroscopy." *Forensic Science International* 211, no. 1-3: 83–100.

Biancolillo, Alessandra, and Federico Marini. 2018 "Chemometric methods for spectroscopy-based pharmaceutical analysis." *Frontiers in Chemistry* 6, 576.

Bolla, Anmole S., Ashwini R. Patel, and Ronny Priefer. 2020. "The silent development of counterfeit medications in developing countries – a systematic review of detection technologies." *International Journal of Pharm*aceutics 587, 119702. doi.org/10.1016/j.ijpharm.2020.119702.

Bovens, M., B. Ahrens, I. Alberink, A. Nordgaard, T. Salonen, and S. Huhtala. 2019. "Chemometrics in forensic chemistry — part I: implications to the forensic workflow." *Forensic Science International* 301, 82–90. doi.org/10.1016/j.forsciint.2019.05.030.

Brito, Joao B. G., Guilherme B. Bucco, Danielle K John, Marco F. Ferrao, Rafael S. Ortiz, Kristiane C. Mariotti, and Michel J. Anzanello. 2020. "Wavenumber selection based on singular value decomposition for sample classification." *Forensic Science International (Online)* 309. dx.doi.org/10.1016/j.forsciint.2020.110191

Bugay, David E. 2001. "Characterization of the solid-state: spectroscopic techniques." *Advanced Drug Delivery Reviews* 48, no. 1: 43–65. doi.org/10.1016/S0169-409X(01)00101-6

Chan, K. L. Andrew, Stephen V. Hammond, and Sergei G. Kazarian. 2003. "Applications of attenuated total reflection infrared spectroscopic imaging to pharmaceutical formulations." *Analytical Chemistry* 75, no. 9: 2140–2146. doi.org/10.1021/ac026456b

Chen, Hui, Zan Lin and Chao Tan. 2018. "Nondestructive discrimination of pharmaceutical preparations using near-infrared spectroscopy and partial least-squares discriminant analysis." *Analytical Letters* 51, no. 4: 564–574. doi.org/10.1080/00032719.2017.1339070

Chen, Hui, Chao Tan, and Zan Lin. 2019. "Express detection of expired drugs based on near-infrared spectroscopy and chemometrics: a feasibility study." *Spectrochim Acta A Mol Biomol Spectrosc.* 5, no. 220: 117153. doi.org/10.1016/j.saa.2019.117153

Chen, Hui, Zan Lin, and Chao Tan. 2020. "Application of near-infrared spectroscopy and class-modeling to antibiotic authentication." *Analytical Biochemistry* 590, 113514. doi.org/10.1016/j.ab.2019.113514.

Cheng, Cungui, Jia Liu, Hong Wang, and Wei Xiong. 2010. "Infrared spectroscopic studies of Chinese medicines." *Applied Spectroscopy Reviews* 45, no. 3: 165–176. doi.org/10.1080/05704920903574256.

Ciza, P. H., P.-Y. Sacrea, C. Waffoa, L. Coïca, H. Avohoua, J. K. Mbinzeb, R. Ngonoc, R. D. Marinia, Ph. Huberta and E. Ziemonsa. 2019. "Comparing the qualitative performances of handheld NIR and Raman spectrophotometers for the detection of falsified pharmaceutical products." *Talanta* 202, 469–476. doi.org/10.1016/j.talanta.2019.04.049

Cooper, Emma. 2020. "COVID-19 and the counterfeit drug crisis." Pf Media. Accessed on November 01 2021. pharmafield.co.uk/opinion/covid-19-and-the-counterfeit-drug-crisis/

Custers, Deborah, Tim Cauwenbergh, Jean-Luc Bothy, Patricia Courselle, Jacques O. De Beer, Sandra Apers, and Eric Deconinck. 2015. "ATR-FTIR spectroscopy and chemometrics: an interesting tool to discriminate and characterize counterfeit medicines." *Journal of Pharmaceutical and Biomedical Analysis* 112, 181–189. doi. org/10.1016/j.jpba.2014.11.007

Custers, Deborah, Suzanne Vandemoortele, Jean-Luc Bothy, Jacques O. De Beer, Patricia Courselle, Sandra Apers, and Eric Deconinck. 2016. "Physical profiling and IR spectroscopy: simple and effective methods to discriminate between genuine and counterfeit samples of Viagra® and Cialis®." *Drug Test. Analysis* 8, 378–387. doi.org/ 10.1002/dta.1813

de Peinder, Peter, Marjo J. Vredenbregt, Tom Visser and Dries de Kaste. 2008. "Detection of Lipitor counterfeits: a comparison of NIR and Raman spectroscopy in combination with chemometrics." *Journal of Pharmaceutical and Biomedical Analysis* 47, no. 4-5: 688–694.

Deconinck, Eric, Sacré, Pierre. Y., Coomans, D., and Jacques De Beer. 2012. "Classification trees based on infrared spectroscopic data to discriminate between genuine and counterfeit medicines." *Journal of Pharmaceutical and Biomedical Analysis* 57, 68–75. doi.org/10.1016/j.jpba.2011.06.036

Dégardin, Klara, Aurelie Guillemain, Nicole V. Guerreiro, and Yves Roggo. 2016. "Near-infrared spectroscopy for counterfeit detection using a large database of pharmaceutical tablets." *Journal of Pharmaceutical and Biomedical Analysis* 128, 89–97. doi.org/ 10.1016/j.jpba.2016.05.004

Dégardin, Klara, Aurelie Guillemain, Philippe Klespe, Florine Hindelang, Raphael Zurbach, and Yves Roggo. 2018. Packaging analysis of counterfeit medicines. *Forensic Science International* 291, 144–157. doi.org/10.1016/j.forsciint.2018.08.023

Dégardin, Klara, Marguerite Jamet, Aurelie Guillemain, and Tobias Mohn. 2019. "Authentication of pharmaceutical vials." *Talanta* 198, 487–500. doi.org/10.1016/j.talanta.2019.01.121

Dorlo, Thomas. P., Teunis A. Eggelte, Peter J. de Vries, and Jos H. Beijnen. 2012. "Characterization and identification of suspected counterfeit miltefosine capsules." *The Analyst* 137, no. 5: 1265–1274. doi.org/10.1039/c2an15641e

Dowell, Floyd E., Elizabeth B. Maghirang, Facundo M. Fernández, Paul N. Newton and Michael D. Green. 2008. "Detecting counterfeit antimalarial tablets by near-infrared spectroscopy." *Journal of Pharmaceutical and Biomedical Analysis* 48, no. 3: 1011–1014.

Eady, Matthew, Michael Payne, Steve Sortijas, Ed Bethea, and David Jenkins. 2021 "A low-cost and portable near-infrared spectrometer using open-source multivariate data analysis software for rapid discriminatory quality assessment of medroxyprogesterone acetate injectables." *Spectrochimica Acta Part A: Molecular and Biomolecular Spectroscopy* 259, 119917. doi.org/10.1016/j.saa.2021.119917

Farouk, Faten, Bahia A. Moussa and Hassan M. E. S. Azzazy. 2011. "Fourier transform infrared spectroscopy for in-process inspection, counterfeit detection and quality control of anti-diabetic drugs." *Spectroscopy* 26, 297–309.

Ghanem, Naira. 2019. "Substandard and falsified medicines: global and local efforts to address a growing problem." Pharmaceutical Journal Accessed November 25 2021. https://pharmaceutical-journal.com/article/research/substandard-and-falsified-medici nes-global-and-local-efforts-to-address-a-growing-problem

Hattori, Yusuke, Yurie Seko, Jomjai Peerapattana, Kuniko Otsuka, Tomoaki Sakamoto and Makoto Otsuka. 2018. "Rapid identification of oral solid dosage forms of counterfeit pharmaceuticals by discrimination using near-infrared spectroscopy." *Bio-medical Materials and Engineering* 29, no. 1: 1–14.

Jamwal, Rahul, Amit, Shivani Kumari, Sushma Sharma, Simon Kelly, Andrew Cannavan, and Dileep Kumar Singh. 2021. "Recent trends in the use of FTIR spectroscopy integrated with chemometrics for the detection of edible oil adulteration." *Vibrational Spectroscopy* 113, 103222. doi.org/10.1016/j.vibspec.2021.103222.

Janvier, Steven, Bart De Spiegeleer, Celine Vanhee, and Eric Deconinck. 2018. "Falsification of biotechnology drugs: current dangers and/or future disasters?" *Journal of Pharmaceutical and Biomedical Analysis* 161, 175–191.

Jarrett, Stephen, Taufik Wilmansyah, Yudha Bramanti, Hikmat Alitamsar, Drajat Alamsyah, Komarapuram R. Krishnamurthy, Lingjiang Yang, and Sonia Pagliusi. 2020. "The role of manufacturers in the implementation of global traceability standards in the supply chain to combat vaccine counterfeiting and enhance safety monitoring" *Vaccine* 38, no. 52: 8318–8325.

Kahmann, A, M. J. Anzanello, F. S. Fogliatto, W. A. Chaovalitwongse, M. C. A. Marcelo, M. F. Ferrão, R. S. Ortiz, and K. C. Mariotti. 2018. "Interval importance index to select relevant ATR-FTIR wavenumber intervals for falsified drug classification." *Journal of Pharmaceutical and Biomedical Analysis* 158, 494–503. doi.org/10.1016/j.jpba.2018.06.046

Khan, Ahmed N., Roop K. Khar, and P. V. Ajayakumar. 2016. "Diffuse reflectance near-infrared-chemometric methods development and validation of amoxicillin capsule formulations." *Journal of Pharmacy & Bioallied Sciences* 8, no. 2: 152–160. doi.org/10.4103/0975-7406.175973

Koech, Lilian C., Beatrice N Irungu, Margeret M. Ng'ang'a, Joyce M. Ondicho, and Lucia K. Keter. 2020. "Quality and brands of amoxicillin formulations in Nairobi, Kenya." BioMed Research International 7091276. doi.org/10.1155/2020/7091278

Kovacs, Stephanie, Stephen E. Hawes, Stephen N. Maley, Emily Mosites, Ling Wong, and Andy Stergachis. 2014. "Technologies for detecting falsified and substandard drugs in low and middle-income countries." *PLoS ONE* 9, no. 3: e90601. doi.org/10.1371/journal.pone.0090601

Krakowska, Barbara, Deborah Custers, Eric Deconinck, and Michal Daszykowski. 2016. "Chemometrics and the identification of counterfeit medicines—a review." *Journal of Pharmaceutical and Biomedical Analysis* 127, 112–122. doi.org/10.1016/j.jpba.2016.04.016.

Laasonen, Magali, Tuulikki Harmia-Pulkkinen, Christine Simard, Markku Räsänen, and Heikki Vuorela. 2004. "Determination of the thickness of plastic sheets used in blister packaging by near-infrared spectroscopy: development and validation of the method." *European Journal of Pharmaceutical Sciences* 21, no. 4: 493–500. doi.org/10.1016/j.ejps.2003.11.011.

Lawson, Graham, Edward Turay, Rachel Armitage, Larry Goodyer, and Sangeeta Tanna. 2014. "Is it what it says on the packet? ATR FTIR provides a rapid answer to counterfeit tablet formulations." *British Global and Travel Health Association Journal*, 23, 55–57.

Lawson, Graham, John Ogwu, and Sangeeta Tanna. 2018. "Quantitative screening of the pharmaceutical ingredient for the rapid identification of substandard and falsified medicines using reflectance infrared spectroscopy." *PLoS ONE* 13, no 8: e0202059. doi.org/10.1371/journal.pone.0202059.

Li, Lian, Hengchang Zang, Jun Li, Dejun Chen, Tao Li, and Fengshan Wang. 2014. "Identification of anisodamine tablets by Raman and near-infrared spectroscopy with chemometrics." *Spectrochimica acta. Part A, Molecular and Biomolecular Spectroscopy* 127, 91–97. doi.org/10.1016/j.saa.2014.02.022.

Lohumi, Santosh, Sangdae Lee, Hoonsoo Lee, and Byoung Kwan Cho. 2015. "A review of vibrational spectroscopic techniques for the detection of food authenticity and adulteration." *Trends in Food Science and Technology* 46, 85–96. doi.org/10.1016/j.tifs.2015.06.003.

Lopes, Marta B. and Jean-Claude Wolff. 2009. "Investigation into classification/sourcing of suspect counterfeit Heptodin tablets by near-infrared chemical imaging." *Analytica Chimica Acta* 633, no. 1: 149–55.

Mackey, Tim K. 2018. "Prevalence of substandard and falsified essential medicines:stillanincompletepicture."*JAMANetworkOpen*1,no.4:e181685.doi./10.1001/jamanetworkopen.2018.1685

Mariotti, Kristiane, Rafael S. Ortiz, Daniele Z. Souza, Thayse C. Mileski, Pedro E. Fröehlich, and Renata P. Limberger. 2013. "Trends in counterfeits amphetamine-type stimulants after its prohibition in Brazil." *Forensic Science International* 229 no. 1-3: 23–26. doi.org/10.1016/j.forsciint.2013.03.026

Marson, Breno M., Raquel D. Vilhena, Camilla R. D. Madeira, Flavia L. D. Pontes, Mario S. Piantavini, and Roberto Pontarolo. 2016. "Simultaneous quantification of artesunate and mefloquine in fixed-dose combination tablets by multivariate calibration with middle infrared spectroscopy and partial least squares regression." *Malaria Journal* 15, 109. doi.org/10.1186/s12936-016-1157-1

Mbinze, Jérémie Kindenge, Pierre Sacré, Achille Yemoa, J Mavar Tayey Mbay, Védaste Habyalimana, Nicodème Kalenda, Philippe Hubert, Roland Djang'eing'a Marini and Éric Ziemons. 2015. "Development, validation and comparison of NIR and Raman methods for the identification and assay of poor-quality oral quinine drops." *Journal of Pharmaceutical and Biomedical Analysis* 111, 21–27. doi.org/10.1016/j.jpba.2015.02.049

Miller, James, and Jane C. Miller. *Statistics and Chemometrics for Analytical Chemistry.* Harlow, Pearson Education, 2018.

Mittal, Manya, Sharma Kritika, and Rathore Anurag. 2021. "Checking counterfeiting of pharmaceutical products by attenuated total reflection mid-infrared spectroscopy". *Spectrochim Acta Part A Molecular and Biomolecular Spectroscopy.* 5, no. 255: 119710. doi.org/10.1016/j.saa.2021.119710.

Moffat, Anthony C., Sulaf Assi, and Robert A. Watt. 2010. "Identifying counterfeit medicines using near-infrared spectroscopy." *Journal of Near-Infrared Spectroscopy* 18, no. 1: 1–15. doi.org/10.1255/jnirs.856.

Mwai, Peter. 2020. "How bad is Africa's counterfeit medicine problem?" BBC news. Accessed on October 30 2021. www.bbc.co.uk/news/world-africa-51122898

Neves, Diana B. J., Marcio Talhavini, Jez W. B. Braga, Jorge J. Zacca, and Eloisa D. Caldas. 2017. "Detection of counterfeit Durateston® using Fourier transform infrared spectroscopy and partial least squares – discriminant analysis." *Journal of the Brazilian Chemical Society [online]* 28, no. 07: 1288–1296. doi.org/10.21577/0103-5053.20160293.

Newton, Paul N., Katherine C. Bond and 53 signatories from 20 countries. 2020. "COVID-19 and risks to the supply and quality of tests, drugs, and vaccines." *The Lancet. Global Health* 8, no. 6: e754–e755. doi.org/10.1016/S2214-109X(20)30136-4

Ogwu, John Epoh. "Forensic pharmaceutical analysis of counterfeit medicines." PhD thesis, De Montfort University, Leicester, 2018.

O'Neil, Andrew J., Roger D. Jee, Ged Lee, Andrew Charvill, and Anthony C. Moffat. 2008. "Use of a portable near-infrared spectrometer for the authentication of tablets and the detection of counterfeit versions." *Journal of Near-infrared Spectroscopy* 16, no. 3: 327–33. doi.org/10.1255/jnirs.796.

Ortiz, Rafael S., Kristiane D. Mariotti, Bruna Fank, Renata P. Limberger, Michel J. Anzanello and Paulo Mayorga. 2013. "Counterfeit Cialis and Viagra fingerprinting by ATR-FTIR spectroscopy with chemometry: can the same pharmaceutical powder mixture be used to falsify two medicines?" *Forensic Science International* 226, no. 1-3: 282–289. doi.org/10.1016/j.forsciint.2013.01.043

Pasquini, Celio. 2018. "Near-infrared spectroscopy: a mature analytical technique with new perspectives – a review." *Analytica Chimica Acta* 1026, 8–36. doi.org/10.1016/j.aca.2018.04.004.

Planinšek, Odon, Daniela Planinšek, Anamarija Zega, Matej Breznik, and Stane Srčič. 2006. "Surface analysis of powder binary mixtures with ATR FTIR spectroscopy." *International Journal of Pharmaceutics* 319, no. 1–2: 13–19. doi.org/10.1016/j.ijpharm.2006.03.048.

Rebiere, Hervé, Céline Ghyselinck, Laurent Lempereur and Charlotte Brenier. 2016. "Investigation of the composition of anabolic tablets using near infrared spectroscopy and Raman chemical imaging." *Drug Testing and Analysis* 8, 3-4: 370–377.

Rebiere, Hervé, Pauline Guinot, Denis Chauvey, and Charlotte Brenier. 2017. "Fighting falsified medicines: the analytical approach." *Journal of Pharmaceutical and Biomedical Analysis* 5, no. 142: 286–306. doi.org/10.1016/j.jpba.2017.05.010

Rebiere, Hervé, Y Grange, E Deconinck, Patricia Courselle, Jeleba Acevska, Katerina Brezovska, Jan Maurin, Torgny Rundlöf, Maria J. Portela, L.S. Olsen, C. Offerlé and Maria Bertrand. 2021. "European fingerprint study on omeprazole drug substances using a multi analytical approach and chemometrics as a tool for the discrimination of manufacturing sources." *Journal of Pharmaceutical and Biomedical Analysis* 208, 114444. doi.org/10.1016/j.jpba.2021.114444.

Reich, Gabriele. 2005. "Near-infrared spectroscopy and imaging: basic principles and pharmaceutical applications." *Advanced Drug Delivery Reviews* 57, no. 8: 1109–1143. doi.org/10.1016/j.addr.2005.01.020.

Ricci, Camilla, Charlotte Eliasson, Neil A. Macleod, Paul N. Newton, Pavel Matousek and Sergei G. Kazarian. 2007. "Characterization of genuine and fake artesunate antimalarial tablets using Fourier transform infrared imaging and spatially offset Raman spectroscopy through blister packs." *Analytical and Bioanalytical Chemistry* 389, 1525–1532.

Rodionova, Oxana. Y., Ksenia. S. Balyklova, A.V. Titova, Pomerantsev, Alexey L. Pomerantsev. 2018. "Application of NIR spectroscopy and chemometrics for revealing of the 'high quality fakes' among the medicines." *Forensic Chemistry* 8, 82–89. doi.org/10.1016/j.forc.2018.02.004

Rodionova, Oxana Y., A.V. Titova, N. A. Demkin, Ksenia S. Balyklova, and Alexey L. Pomerantsev. 2019. "Detection of counterfeit and substandard tablets using non-invasive NIR and chemometrics – a conceptual framework for a big screening system." *Talanta* 205, 120150. doi.org/10.1016/j.talanta.2019.120150.

Rodionova, Oxana Y., A.V. Titova, N. A. Demkin, Ksenia S. Balyklova, and Alexey L. Pomerantsev. 2021. "Influence of the quality of capsule shell on the non-invasive monitoring of medicines using Terizidone as an example" *Journal of Pharmaceutical and Biomedical Analysis* 204, 114245. doi.org/10.1016/j.jpba.2021.114245

Roggo, Yves, Pascal Chalus, Lene Maurer, Carmen Lema-Martinez, Aurélie Edmond, and Nadine Jent. 2007. "A review of near-infrared spectroscopy and chemometrics in pharmaceutical technologies." *Journal of Pharmaceutical and Biomedical Analysis* 44, no. 3: 683–700. doi.org/10.1016/j.jpba.2007.03.023.

Roth, Lukas, Kevin B. Biggs, and Daniel K. Bempong. 2019. "Substandard and falsified medicine screening technologies." *AAPS Open* 5, 2. doi.org/10.1186/s41120-019-0031-y.

Sacré, Pierre-Yves, Eric Deconinck, Thomas De Beer, Patricia Courselle, Roy Vancauwenberghe, Patrice Chiap, Jacques Crommen, and Jacques O. De Beer. 2010. "Comparison and combination of spectroscopic techniques for the detection of counterfeit medicines." *Journal of Pharmaceutical and Biomedical Analysis* 53, no. 3: 445–453. doi.org/10.1016/j.jpba.2010.05.012

Said, Mazlina M., Simon Gibbons, Anthony C. Moffat and Mire Zloh. 2011. "Near-infrared spectroscopy (NIRS) and chemometric analysis of Malaysian and UK paracetamol tablets: a spectral database study." *International Journal of Pharmaceutics* 415, no. 1-2: 102–109.

Said, Mazlina M., Simon Gibbons, Anthony Moffat, and Mire Zloh. 2019. "Use of near-infrared spectroscopy and spectral databases to assess the quality of pharmaceutical products and aid characterization of unknown components." *Journal of Near-infrared Spectroscopy* 27, no. 5: 379–390. doi.org/10.1177/0967033519866009.

Salari, Amid, and Richard E. Young. 1998. "Application of attenuated total reflectance FTIR spectroscopy to the analysis of mixtures of pharmaceutical polymorphs." *International Journal of Pharmaceutics* 163, no. 1–2: 157–166. doi.org/10.1016/S0378-5173(97)00378-5.

Salim, Mohammad R., Riyanto T. Widodo, and Mohamed I. Noordin 2021. "Proof-of-concept of detection of counterfeit medicine through polymeric materials analysis of plastics packaging" *Polymers* 13, no. 13: 2185. doi.org/10.3390/polym13132185

Sammons, Helen M., and Imti Choonara. 2017. "Substandard medicines: a greater problem than counterfeit medicines?." *BMJ Paediatrics Open* 1, no. 1(May): e000007. doi.org/10.1136/bmjpo-2017-000007

Scafi, Sergio H. F., and Celio Pasquini. 2001. "Identification of counterfeit drugs using near-infrared spectroscopy." *The Analyst* 126, 2218–2224. doi.org/10.1039/b106744n

Si, Leting, Hongfei Ni, Dongyue Pan, Xin Zhang, Fangfang Xu, Yun Wu, Lewei Bao, Zhenzhong Wang, Wei Xiao, and Yongjiang Wu. 2021. "Nondestructive qualitative and quantitative analysis of yaobitong capsule using near-infrared spectroscopy in tandem with chemometrics." *Spectrochimica Acta – Part A: Molecular and Biomolecular Spectroscopy* 252, 119517. doi.org/10.1016/j.saa.2021.119517.

Srivastava, Kanchan. 2021. "Fake covid vaccines boost the black market for counterfeit medicines" *BMJ* 375, 2754. doi.org/10.1136/bmj.n2754

Storme-Paris, Isabelle, Hervé Rebiere, Myriam Matoga, Corinne Civade, Pierre Bonnet, M. H. Tissier and Pierre Chaminade. 2010. "Challenging near-infrared spectroscopy discriminating ability for counterfeit pharmaceuticals detection." *Analytica Chimica Acta* 658, no. 2: 163–174.

Tanna, Sangeeta and Graham Lawson. 2016. "Medication adherence." In *Analytical Chemistry for Assessing Medication Adherence*, edited by Brian F. Thomas, 1–21. Amsterdam: Elsevier.

Tie, Yaxin, Céline Duchateau, Shana Van de Steene, Corenthin Mees, Kris De Braekeleer, Thomas De Beer, Erwin Adams, and Eric Deconinck. 2020.

"Spectroscopic techniques combined with chemometrics for fast on-site characterization of suspected illegal antimicrobials." *Talanta* 217, 121026. doi.org/ 10.1016/j.talanta.2020.121026.

Tesfaye, Wubshet, Solomon Abrha, Mahipal Sinnollareddy, Bruce Arnold, Andrew Brown, Cynthia Matthew, Victor M. Oguoma, Gregory M. Peterson, and Jackson Thomas. 2020. "How do we combat bogus medicines in the age of the COVID-19 pandemic?" *The American Journal of Tropical Medicine and Hygiene* 103, no. 4: 1360–1363. doi. org/10.4269/ajtmh.20-0903

Venhuis, Bastiaan J., Gijsbert Zomer, Marjo J. Vredenbregt and Dries de Kaste. 2010. "The identification of (-)-trans-tadalafil, tadalafil, and sildenafil in counterfeit Cialis and the optical purity of tadalafil stereoisomers." *Journal of Pharmaceutical and Biomedical Analysis* 51, no. 3: 723–727.

Vickers, Serena, Matthew Bernier, Stephen Zambrzycki, Facundo M. Fernandez, Paul N. Newton, Céline Caillet. 2018. "Field detection devices for screening the quality of medicines: a systematic review." *BMJ Glob. Health* 3, e000725. doi.org/10.1136/ bmjgh-2018-000725

Vredenbregt, Marjo J., Leonore Blok-Tip, Ronald Hoogerbrugge, Dirk M. Barends and Dries de Kaste. 2006. "Screening suspected counterfeit Viagra and imitations of Viagra with near-infrared spectroscopy." *Journal of Pharmaceutical and Biomedical Analysis* 40, no. 4: 840–849.

Waffo Tchounga, C. A., P. Y. Sacre, P. Ciza, R. Ngono, E Ziemons, Ph. Hubert, and R. D. Marini. 2021. "Composition analysis of falsified chloroquine phosphate samples seized during the COVID-19 pandemic." *Journal of Pharmaceutical and Biomedical Analysis*, 194, 113761. doi.org/10.1016/j.jpba.2020.113761

Wang, Wenbo, Matthew D. Keller, Ted Baughman, and Benjamin K. Wilson. 2019. "Evaluating low-cost optical spectrometers for the detection of simulated substandard and falsified medicines." *Applied Spectroscopy* 74, no. 3: 323–333. doi.org/10.1177/ 0003702819877422.

Wilson, Benjamin K, Harprakash Kaur, Elizabeth L. Allan, Anthony Lozama and David Bell. 2017. "A new handhled device for the detection of falsified medicines: demonstration on falsified artemisinin-based therapies from the field." *The American Journal of Tropical Medicine and Hygiene* 96, no. 5: 1117–1123 doi.org/10.4269/ajtmh.16-0904.

World Health Organisation. 2017. "*A Study on the Public Health and Socioeconomic Impact of substandard and falsified medical products.*" Geneva, Switzerland: World Health Organisation.

World Health Organisation. 2018. "*Substandard and falsified medical products.*" Accessed on October 02 2021. www.who.int/news-room/fact-sheets/detail/substandard-and-falsified-medical-products

Xu, Yi, Peng Zhong, Aimin Jiang, Xing Shen, Xiangmei Li, Zhenlin Xu, Yudong Shen, Yuanming Sun, and Hongtao Lei. 2020. "Raman spectroscopy coupled with chemometrics for food authentication: a review." *TrAC – Trends in Analytical Chemistry* 131, 116017. doi.org/10.1016/j.trac.2020.116017.

Zontov, Y. V., K. S. Balyklova, A. V. Titova, O. Y. Rodionova, and A. L. Pomerantsev. 2016. "Chemometric aided NIR portable instrument for rapid assessment of medicine quality." *Journal of Pharmaceutical and Biomedical Analysis* 131, 87–93. doi.org/ 10.1016/j.jpba.2016.06.008

7 Using NMR Spectroscopy to Analyze Counterfeit Medications

*Alina Hoxha[1] and Ronny Priefer[1, *]*
[1]MCPHS University, 179 Longwood Ave, Boston, MA, USA, 02115

CONTENTS

7.1 Introduction 171
7.2 Nuclear Magnetic Resonance (NMR) 172
7.3 Quantitative NMR 173
7.4 Qualitative NMR 180
7.5 Conclusion 185
References 188

7.1 INTRODUCTION

Counterfeit medications are a significant issue for patients worldwide and present as a general public health threat. The World Health Organization (WHO) suggests that approximately 6% of global, commercially available drugs are counterfeit, while the FDA places this closer to 10%. Producing illegal, counterfeit forms of drugs is a "big money-making" industry, with an estimate of $75 billion annually in the United States alone [1]. It also leads to increased morbidity and mortality rates, due to subtherapeutic or adverse side effects. [2] This, in turn, costs global healthcare approximately $32 billion annually. [2]

Counterfeit medications are products that either do not contain the ingredients as listed on the package insert/label, do not have the correct active ingredient, have less of the ingredient(s), or none whatsoever. [3] Other examples of counterfeit drugs are when the formulation contains alternative additives than claimed or being a different drug altogether. This potential of consuming ingredients that are not indicated for a specific treatment can lead to serious health issues. In 2013, aspirin and Pegasys syringes were reported on the European market as adulterated, containing glucose instead of their active ingredients. [3] Herbal supplements are yet another example of counterfeits. Some are marketed as not interacting with medications, having no side effects, and as alternatives to PDE-5 inhibitors. Nevertheless, many of them were found to actually contain PDE-5 inhibitors, such as sildenafil, vardenafil, and tadalafil. [4] Another risk posed by counterfeit drugs is that patients may be taking a formulation that does not possess the correct active pharmaceutical ingredient (API), or none at all. This can lead to the patient having the wrong treatment or regimen, and/or causing toxicities. One such example of a drug not containing the API was in 2012, with the anticancer drug, Avastin. The "medication" contained corn starch and acetone instead of the biologic, bevacizumab. [5] Likewise, when a drug has been altered to have less of the active agent than claimed, patients are underdosed, which can lead to resistance to treatment. This has been observed with antibiotics and antivirals leading to an increase in aggressive infection resistance to pneumonia and HIV treatments. [2]

Counterfeit medications are more prevalent in developing countries, largely due to: (1) lack of appropriate regulations, (2) poor resources, such as evaluating the quality of drugs, (3) shortage of funds to obtain expensive, more regulated drugs, and (4) increased demand for certain treatments. [2] Examples in parts of Africa include antimalarials and medications to treat HIV/AIDS [1], which are widely needed due to the severity and the abundance of these pathogenic diseases within the continent.

Counterfeit drugs are not only a "third-world" problem. In developed countries, such as the US, Canada, and within Europe, incidences have been observed. This is more readily seen with herbal medications, dietary supplements, and lifestyle modification enhancers, which tend to be marketed on false websites. [2]

Currently, there are several different methods to assess for impurities or false claims regarding counterfeit drugs. Separation techniques, such as chromatography are good for detecting impurities in additives, but are not always effective at identifying APIs. Spectrometry tests are good for chemical imaging and include methods based on vibrations, such as IR, UV-Vis, and Raman. Mass spectrometry (MS) is able to detect and quantify APIs based on their mass-to-charge ratio. These technologies are exceptionally useful, albeit all have their own strengths and weaknesses. Nuclear magnetic resonance (NMR) is a very powerful technology and has been shown to be effective at detecting counterfeit medications. Herein, is an overview of studies that have utilized NMR, to either identify or quantify APIs in hopes to combat the counterfeit drug market.

7.2 NUCLEAR MAGNETIC RESONANCE (NMR)

NMR is a spectroscopic method that measures the nuclear spins of atoms and how they interact when presented with electromagnetic radiation. This leads to their resonance and reveals both the chemical and magnetic environment of the various atomic nuclei of a molecule. From the obtained spectra it is possible to determine the precise structure of the molecule being analyzed. There are many different types of NMR; however, ^{1}H-NMR is the most common due to the ^{1}H being the major isotope of hydrogen and having a spin of ½. Other nuclei have also been utilized by NMR, such as ^{13}C, ^{15}N, and ^{19}F due to the fact that they also have a spin of ½.

Within the choices of classic NMR, there exist either high- or low-field NMR. The main difference is not only the cost of upkeep, but also its footprint (it is quite bulky) and the spectral resolution. In one study [6], high-field ^1H-NMR (HF ^1H-NMR) was shown to successfully identify and quantify synthetic cannabinoids in spice herbal blends. In this study, there was no need for a specific standard for quantification. In the same study, low-field ^1H-NMR (LF ^1H-NMR) showed to be a useful method for rapid screening, due to having quick sample preparation and rapid spectral recording. [6]

Additional experiments that require high resolution are 2D-NMRs, which provide more clarity and/or sample identity. One study utilized 2D diffusion-ordered spectroscopy (DOSY) NMR to give a precise signature of genuine Cialis®, and was able to identify seven different imitations of the drug being counterfeited. [7]

Time domain NMR (TD-NMR) has been recently employed to detect counterfeit biologic drugs. It is a rapid technique and does not require the sample to be dissolved in a deuterated solvent. It analyzes the water content of lyophilized proteins as well as their aggregation in solution. Thus, it can be used for mixes of drugs in solid states. Some drugs that are susceptible to counterfeiting and were able to be detected by TD-NMR were Avastin, rabies vaccines, erythropoietin, and monoclonal antibodies (mAbs), among others. [8] Additionally, TD-NMR was employed to determine the structures of four different mAbs, and whether the drugs were diluted.

Another approach to test biologic pharmaceuticals is via solid-state NMR (ssNMR). Looking at which polymorph of a solid-state drug, such as Clopidogrel, the final product was successfully analyzed. [9] The Clopidogrel tablets were first homogenized with no further need for sample preparation prior to running, which made it easier, faster, and less costly with regards to preparation time. [9]

7.3 QUANTITATIVE NMR

Pezza *et al.*, used ^1H-NMR to determine the authenticity of 16 samples of six different anabolic steroids available as injectables, tablets, and/or capsules sold commercially in Brazil. [10] The samples were obtained by the Brazilian federal police and included two injectables and two tablets of stanozolol (STZ), two tablets and two capsules of oxandrolone (OXA), four injectables of testosterone propionate (TPR), and four injectables of nandrolone decanoate (NDE). They initially ran a qualitative ^1H-NMR on a 600 MHz instrument to obtain quick results on whether the samples contained any APIs. Of the 16 samples, two each of TPR and NDE contained no active agents. Subsequently, quantitative ^1H-NMR studies were conducted on the remaining 12 samples employing three different methods: (1) direct, (2) analytical curve, and (3) comparative. Results from these methods were compared by setting a critical value of the t-test at 4.303, with a

95% confidence interval. Not exceeding this critical value, indicated no significant difference between the methods. [10]

The direct determination method used 2 mg/mL of dimethyl sulfone ($DMSO_2$) as an internal reference which was added to the sampled anabolic steroids. [10] Using the same 600 MHz NMR, a chemical shift signal at 3.00 ppm for $DMSO_2$ was recorded. No other proton signals interfered or overlapped with this peak. ^{1}H-NMR, with add $DMSO_2$, was run on the 12 samples that showed presence of the API after qualitative analysis. One of each sample of injectable STZ, TPR, and NDE showed the presence of the API in a lower amount than labeled, as well as containing impurities. This method's key advantage was not needing analyte standards, which often are not readily available, as well as being a time-efficient experiment. Moreover, this method was a practical and nondestructive approach requiring minimal sample preparation steps. [10]

The analytical curve method (aka calibration curve) was also evaluated for its suitability for quantitative determination of the APIs in the samples. [10] ^{1}H-NMR spectra were run on different concentrations of standard samples of the six anabolic steroids (purchased from Sigma-Aldrich), as well as on standard peanut oil and benzyl alcohol, which are often used as excipients during manufacturing. The concentration range of the standard samples was 1.0–16.0 mg/mL. Correlation coefficients (r) were compared to the Brazilian National Health Surveillance Agency (Agência Nacional de Vigilância Sanitária; ANVISA) and the International Conference on Harmonization (ICH). Two intra-day and one inter-day precision experiments were also performed on all twelve samples to determine the repeatability of the method within the same day and on different days. Additionally, the standard solution additions were used to fortify the following samples: STZ tablet, OXA capsule, and injectables of TPR and NDE. The analytical method was studied for its robustness by seeing if there were a difference depending on NMR tubes, specifically 3 mm and 5 mm tubes, as well as comparing two types of data processing software. Ultimately, only the STZ tablets and injectable TPR samples were fully analyzed. It was concluded that this method was not influenced by operational variations. Additionally, similar to the direct determination method, it was determined that one of each sample of injectable STZ, TPR, and NDE contained less API than labeled and had impurities of peanut oil and benzyl alcohol. Comparing the analytical curve to the direct determination method, no significant differences were observed. However, the analytical curve method needed several analyte standards, more time, and was more costly. [10]

From the comparative methods – aka high-performance liquid chromatography, (HPLC) – three kinds of literature were used as a reference, to determine NDE[10/39], TPR[10/39], STZ[10/40] and American Pharmacopoeia for determining OXA polarity profile. [10] Different conditions, such as mobile and stationary phases, scan speed, and wavelength were different in all of the six unique anabolic steroid standards. The comparative method did not show a significant difference between the prior two approaches regarding the quantification of the anabolic steroids' samples; however, it had challenges associated with sample preparation. Generally, of the three methods employed, none showed analytical superiority, however, the direct determination approach was the most cost and time-efficient. [10]

McEwen *et al.,* used ^{1}H-NMR in combination with ^{19}F-NMR and ^{13}C-NMR, as well as LC-MS analysis to screen for counterfeit corticosteroids in creams and ointments. [11] The NMR spectra of the three nuclei were obtained using the aforementioned direct method. Initially, quantitative NMR spectroscopy was used to identify and quantify 16 types of corticosteroids in licensed products, to validate the accuracy of the methods. The samples had a known API, concentrations between 0.05%–0.1%, and were obtained directly from pharmaceutical companies in Germany, Sweden, Switzerland, Pakistan, and the UK. The samples included dexamethasone 21 acetate, triamcinolone acetonide, mometasone furoate, methylprednisolone, clobetasone butyrate, fluticasone, fluprednidene, fluorometholone, fluorocortolone, fluocinolone acetonide, flumethasone, fluclorolone acetonide, diflorasone diacetate, desoximethasone, betamethasone acetate, and betamethasone 17-valerate. [11]

The quantitative ^{1}H-NMR spectroscopy was run using a 300 MHz NMR. The internal standard used as reference was 2,4-dinitrofluorobenzene which has two signals, 8.6 ppm and 9.0 ppm. [11] The spectra were obtained from different sample preparations (triplication) to assess the relative standard deviation (RSD), which was set at 15%. It was observed that the spectra for the 16 samples were crowded with other chemical signals, however, some signals for both protons H^2 and H^4 were different in different creams/ointments. Ultimately, ^{1}H-NMR was sufficient to identify and quantify the corticosteroids, but due to the variations in the spectra, ^{19}F-NMR analysis was employed to verify the results. [11]

Because most corticosteroids from the samples possessed fluorine atoms, the ^{19}F-NMR spectra revealed one or more signals. [11] Therefore, ^{19}F-NMR was used in combination with ^{1}H-NMR to confirm the identity of the samples. The aformentioned internal standard had a signal at 107 ppm and could be used to quantify the API. The authors reported that ^{19}F-NMR gave similar results to that obtained with ^{1}H-NMR with an RSD within the range of <15%. [11]

After analyzing the licensed samples with known API and standardized concentrations, ten over-the-counter (OTC) creams/ointments obtained from health shops in China, that allegedly were not considered medicine, were tested for corticosteroids. [11] The OTC products included Bai Fu Kang Unguent ointment, Shenzhen 999 cream, Alanoferm cream, Cadramine-V cream, ointment of coca body butter, hemp body butter, Miafrid pure night cream, secuvie body repair body lotion, XM™-FOCUS cream, and Z™- skin repair cream. ^{1}H-NMR was run on all ten samples. Two were identified and quantified for the presence of a corticosteroid. One of the samples, Shenzhen 999 cream, had declared the presence of the corticosteroid dexamethasone 21-acetate 0.075%. ^{19}F-NMR was employed to confirm these claims and it was seen that this sample had a concentration of 0.09% of the corticosteroid with an RSD of 5%. Bai Fu Kang Unguent ointment had no declared corticosteroid; however, both ^{1}H-NMR and ^{19}F-NMR determined the presence of corticosteroid, although of which, was unknown. Using LC-MS the identification of the corticosteroid was determined to be 0.11% triamcinolone acetonide. [11]

Another study that utilized ^{1}H-NMR spectroscopy, via the direct determination method, was by Shi et al.,, who analyzed six reference standards of angiotensin-converting-enzyme (ACE) inhibitors for their purity. [12] The samples included imidapril hydrochloride, benazepril hydrochloride, lisinopril, enalapril maleate, quinapril hydrochloride, and captopril, which were obtained from the National Institute for Food and Drug Control in China. The internal standard used for reference (Sigma-Aldrich in Switzerland) was 0.02 mol/L maleic acid. For the captopril sample, an extra step utilizing HPLC was employed to isolate the ACE inhibitor from the tablet. Quantitative ^{1}H-NMR was run on all samples, with the addition of repeatability and stability being tested by performing six runs for each sample on the same day at two-hour intervals with the same analyst. The results were compared to those determined by HPLC. Overall, the ^{1}H-NMR method showed to be a rapid and useful way to determine the purity of the ACE inhibitors. [12]

Dagnino et al., also employed the direct determination method with ^{1}H-NMR on samples of five different drugs of different brands with a dose range of 0.25–750 mg. [13] The samples included 11 tablets of sertraline, four tablets of alprazolam, three capsules of vitamin D3, five tablets of enalapril maleate, and two tablets of paracetamol, obtained from various pharmacies in Brazil. ^{1}H-NMRs from a 500 MHz instrument were run with the internal standard of DMSO-d$_6$. After the quantitative analysis, it was observed that two samples of alprazolam tablets were of different brands from the same industrial group, while the other two had a lower concentration of the API than indicated. Furthermore, all samples of vitamin D3 capsules had a lower concentration of their API than claimed. The ten samples of sertraline, as well as samples of enalapril maleate, also showed high deviations in the quantity of the API. Finally, both paracetamol tablets were found to be contaminated with higher levels than allowed of para-aminophenol. [13]

Similarly, the ^{1}H-NMR direct determination method was also used by Rebiere et al., in their fingerprint study on 28 authentic samples of omeprazole tablets from eleven legitimate manufacturers authorized in Europe. [14] This study aimed to determine the best method that would distinguish the different manufacturers and establish fingerprints as a future reference for detecting counterfeit omeprazole manufactured in poor quality. In addition to NMR, analysis included HPLC, GC-MS, IR, and X-ray powder diffractometry (XRPD). Two chemometric methods were also used as tools to aid in the fingerprinting process. These methods included principal component analysis (PCA) plots and hierarchical clustering analysis (HCA) dendrograms. The samples were obtained from different national markets in Europe and were all documented and traceable. Two of the samples contained omeprazole magnesium while the rest were the free-acid, omeprazole. ^{1}H-NMR studies were performed on a 600 MHz instrument, and the internal reference used was 0.6 mL dimethyl sulfoxide-d6. After performing the quantitative ^{1}H-NMR, the two samples containing the salt form of omeprazole were confirmed and not studied further. The presence of propanol, acetone, dichloromethane, and dibutyl phthalate were noted in the remaining 26 samples. PCA was performed, reflecting eight different clusters, with samples from single manufacturers. An HCA dendrogram was also plotted and confirmed the analysis from the PCA. Out of 11 different manufacturers, the use of quantitative ^{1}H-NMR, in combination with PCA and HCA, led to fingerprinting products of eight of them. [14]

Another approach for fingerprinting the origin of APIs was studied by Remaud et al.,. [15] In their study, they used ^{2}H and single-pulse ^{13}C-NMR to identify and quantify the components of 20 samples of commercial ibuprofen obtained in Belgium, France, Portugal, UK, and the US. Additionally, they used the intense nuclei enhanced polarization transfer (INEPT) ^{13}C-NMR to analyze 11 samples of commercial naproxen purchased from pharmacies in China, France, Israel, and Russia. Initially, they ran a quick qualitative ^{1}H-NMR to check the purity of the products. Subsequently, quantitative ^{2}H-NMR and single-pulse ^{13}C-NMR on the ibuprofen samples and ^{13}C-INEPT-NMR on naproxen samples were performed. The ^{2}H-NMR was recorded by a 75 MHz instrument, while ^{13}C and ^{13}C-INEPT were recorded using a 100 MHz. The parameters for INEPT included offsetting of the spectra shifts ranging from 4 ppm for ^{1}H and 100 ppm for ^{13}C isotopes. For the quantitative analysis, 300 mg of ibuprofen was dissolved in 600 μL acetone-d$_6$, and 200 mg of naproxen was dissolved in 1000μL of acetone-d$_6$/DMSO 7/3 v/v. The spectra were plotted similarly to Rebiere [14] by PCA. Analysis of all samples of ibuprofen and naproxen demonstrated the ability to fingerprint and identify the samples' origin. The authors concluded that multinuclear NMR spectroscopy is a promising tool to be used for the detection of counterfeit drugs, specifically for tracking, from starting material to final deliverable. [15]

Akoka et al., did a similar study on ibuprofen using quantitative single-pulse ^{13}C-NMR, distortionless enhancement by polarization transfer (DEPT), and ^{13}C-INEPT measurements, with future plans for use to detect counterfeit drugs and track their origins. [16] Thirteen samples of commercial ibuprofen were obtained from the US, UK, Belgium, France, and Portugal. After the qualitative ^{1}H-NMR, quantitative ^{13}C-NMR on a 600 MHz instrument was performed. In total only 232 scans for single-pulse, and 64 for DEPT and INEPT each, were run. All samples were quantified and fingerprinted to their origin. [16]

Malet-Martino et al., employed Raman and NMR in their analysis of eight formulations of 10 mg Tadalafil (Cialis). [7] One of the samples included the genuine Cialis®, obtained from Eli Lilly Laboratories in Germany, two samples were purchased in Syria, four via the internet from India, and one from China. These samples were analyzed initially by Raman spectroscopy to check for the presence of the API as well as for excipients. The spectra from genuine Cialis® were used as a reference for the seven samples. The API was present in all samples purchased from India and Syria; however, the one from China was void of any API. These conclusions were verified using ^{1}H-^{1}H-2D-DOSY-NMR spectroscopy as well as LC-MS analyses. The samples were prepared by dissolving in an 8:2 mix of CD$_3$CN and D$_2$O and run using a 500 MHz instrument. The China sample underwent additional chromatographic purification, followed by direct determination method with ^{1}H-NMR using maleic acid as the internal reference. The sample contained two APIs quantified to be 34.3±0.2 mg of vardenafil and 7.2±0.1 mg of homosildenafil. [7]

Lachenmeier et al., used a standardless methodology referred to as pulse length-based concentration determination (PULCON) with NMR to obtain the purity of reference materials, including ibandronic acid, amantadine, ambroxol, and lercanidipine. [17] They compared the results of their analysis with the direct determination method and with the reference values from HPLC analysis. For the NMR analysis via PULCON and the direct determination method, a 400 MHz NMR was employed. The direct determination method used two internal standards, 3,5-dinitrobenzoic acid and 2,3,4,5-tetracloronitrobenzene. [17]

The PULCON analysis utilized a reference standard from an electronic reference to access in vivo concentrations (ERETIC). [17] The ERETIC factor for the chloroform solutions were determined by adding ethylbenzene in $CDCl_3$, while the factor for the aqueous solutions was determined by adding citric acid in D_2O. Only a single calibration with the PULCON method was needed as it could be reused for multiple samples over an extended period. The reference samples analyzed were as follows: one sample each of ibandronic acid, amantadine HCl, ambroxol HCl, and lercanidipine, and four samples of medicinal products of ibandronic acid. Both types of NMR methods showed satisfactory purity values of all samples of ibandronic acid. The purity using the direct determination method of ^{1}H-NMR was 93.6%–96%, whereas with PULCON it was determined to be 94.8%–99.6%. A t-test showed no significant difference between the two methods and the values agreed with sample specifications as well as with the reference values from HPLC. The authors also utilized the analysis of the three reference samples of amantadine HCl, ambroxol HCl, and lercanidipine to confirm the performance of the PULCON method. They concluded that the PULCON method was effective in checking the specifications of commercial products. Additionally, it had the advantage, compared to direct method ^{1}H-NMR, as to not needing multiple calibrations of the samples being evaluated. [17]

Abraham et al., utilized another unique method of NMR spectroscopy, known as TD-NMR to analyze for counterfeit drugs. [8] This method did not require the dissolving of samples in a deuterated solvent. Abraham determined the structures of four different monoclonal antibodies, CM-1, CM-2, CM-3, and CM-4 (Selleck Chemicals), and whether the drugs were diluted. They ran TD-NMR on samples at an initial concentration of 5 mg/mL. By changing the concentration of the buffer mixed with the antibodies, it was possible to create thirteen variation sets of three buffered solutions. The relaxation times changed depending of the combination employed. It was observed that six samples had no API, and seven of them had lower concentrations than declared. [8]

Casagrande et al.,. analyzed eight monoclonal antibodies samples found on the Swiss market. [18] They prepared the samples by adding 10% D_2O and ran a 600 MHz ^{1}H-NMR via the direct determination method. The internal reference used was a typical monoclonal antibody from the Roche internal spectral database. Two of the tested samples using showed the presence of excipients and no API. One contained hydroxyethyl starch and another benzyl alcohol. Three samples had low molecular weight API, while another had methylprednisolone, diclofenac, or gemcitabine. Additionally, three samples had the wrong biological altogether; being insulin, human serum albumin (HSA), or gelatin. [18]

Some other studies utilizing NMR to detect counterfeit drugs, were from Jin-Lee [19], Maruyama [20], and Reepmeyer [21]. They employed ^{1}H-NMR and ^{13}C-NMR to identify and quantify adulterants in counterfeit drugs, such as in dietary supplements, Saposhnikoviae radix (SA) herbal supplements, and OTC herbal aphrodisiacs, respectively. Similarly, Remaud et al., [22] utilized ^{13}C-NMR to identify and quantify the contents of counterfeit aspirin and paracetamol.

Additional studies that utilized quantitative NMR via an analytical method were by Holzgrabe [23], Malet-Martino [24], Schramek [25], and Kang [26]. Holzgrabe et al., utilized 500 MHz ^{1}H- and ^{13}C-NMR to analyze a sachet of 2.617 g anti-rheumatic drug purchased from Vietnam. [23] They found 33% acetaminophen, 10% sulfamethoxazole, 1.6% indomethacin, and <1% trimethoprim. Similarly, Malet-Martino et al., studied nine samples of dietary supplements (DS) from South Europe claiming to contain a mix of herbs and sugars. [24] By utilizing a 500 MHz ^{1}H-NMR, four of the samples were identified to be adulterated with the sildenafil analogue, propoxylphenyl-thiohydroxyhomosildenafil (PP-THHS), one had thiosildenafil (THIO), two had sildenafil (SILD) and tetrahydropalmatine (THP), one had phentolamine (PHE), and one had osthole (OST). [24] Schramek et al., utilized the analytical method of ^{1}H-NMR to quantify capsule samples of DS purchased from Germany. [25] Using 2D-^{1}H-NMR-COSY, it was determined that the samples contained two analogues of sildenafil, piperazinafil, and isopiperazinonafil in equal amounts. [25] Kang et al., similarly analyzed 23 samples of weight-loss compounds purchased from South Korea, by utilizing the analytical curve method on a 600 MHz via ^{1}H- and ^{13}C-NMR. [26] They found that five out of 23 samples were adulterated; three of them contained levothyroxine, and one each of sennoside A, sennoside B, and phenolphthaleine. In addition, they found three unknown adulterants in one of the samples. [26]

Some other studies that analyzed DS utilizing quantitative NMR were by Malet-Martino [27], Pauli [28], Lachenmeier [29], and Hakkarainen [30]. Malet-Martino et al., utilized a 500 MHz ^{1}H-NMR to analyze 150 samples of sexual enhancement DS purchased online in France. [27] They found sildenafil in 27 samples, tadalafil in 19, vardenafil in two, and flibanserin in four. [28] Similarly, Pauli et al., utilized ^{1}H-NMR on a 400 MHz instrument to analyze turmeric-containing DS samples purchased in Italy. [28] They found 4.8%–7.4% of adulterant synthetic curcumin in the samples, which had been causing acute nonviral cholestatic hepatitis in 21 individuals in the country. [28] Lachenmeier et al., studied 16 DS samples purchased online, utilizing 400 MHz ^{1}H-NMR. [29] They found sibutramine in three samples, mesterolone in two as well as oxymetholone, monacolin K, vinpocetine, evodiamine, caffeine and testosterone propionate each in one sample. [29] Another study by Hakkarainen *et al.*, studied three samples of grapefruit seed extract (GSE) purchased online in Sweden, by using the direct determination method of NMR. [30] They utilized ^{1}H- and ^{13}C-NMR on 300 MHz and 600 MHz broadband probe, and on 600 MHz cryogenic probe concurrently. They observed that all three samples had benzethonium in addition to glycerol and water. No authentic GSE was found in any of the samples.

Bogun *et al.*, utilized direct determination method to analyze liquid methamphetamine samples obtained from clandestine labs in New Zealand. [31] They ran the analysis on a 40 MHz ^{1}H- and ^{31}P. They were able to identify and quantify between phosphorus-containing acids or the equivalent basic anion. [31] Similarly, Kaur *et al.*, utilized ^{1}H- and ^{31}P-NMR on a 400 MHz instrument to analyze three samples of blister packs of miltefosine purchased in Bangladesh. [32] They observed that the samples had no APIs present. [32]

Quantitative studies are quite useful (see Table 7.1), however the actual practicality in the field is limited because they require the utilization of high-field NMR, which often have restrictions regarding cost and maintenance; particularly cryogenic exchange. Often, a qualitative analysis of a sample is sufficient to determine whether the API is present. Thus, low-field and/or benchtop NMR have become increasingly popular.

7.4   QUALITATIVE NMR

Balayssac *et al.*, used low-field benchtop ^{1}H-NMR to analyze counterfeit weight-loss DS purchased online. [33] Initially, they qualified and quantified 40 samples of DS and categorized them to build statistical models. The samples were grouped as nonadulterated, sibutramine-adulterated, phenolphthalein-adulterated, or adulterated with both sibutramine and phenolphthalein. Another 13 samples were purchased from the internet and analyzed to compare with the previous 40 samples. The samples were dissolved in 1 mL of deuterated methanol and a LF-^{1}H-NMR was run on a 60 MHz machine. Three of the samples were phenolphthalein-adulterated and another three were sibutramine-adulterated. The remaining seven were classified as non-adulterated, not containing contaminants. [33]

Pagès *et al.*, utilized benchtop cryogen-free LF-^{1}H-NMR, to detect and identify adulterants in sexual enhancement and weight-loss DS. [34] Eleven sexual enhancement and five weight-loss samples were purchased from the internet claimed to be "100% natural". The samples were initially run on a 500 MHz HF-^{1}H-NMR. The adulterants, sildenafil and tadalafil, were identified in ten of the sexual enhancement and four weight-loss samples. These were also detected using a 60 MHz benchtop instrument, validating the accuracy of LF-benchtop NMR in detecting counterfeit drugs. [34]

TABLE 7.1
Applications of qNMR spectroscopy for the analysis of counterfeit drugs

Technique/Method	Drug	Location	NMR (MHz)	Nuclei/Experiment	Adulterant	Ref
Direct	Anabolic Steroids	Brazil	600	^1H	No API/ Less API/ peanut oil, benzyl alcohol	[10]
	OTC creams/ointments with no API	Germany, Sweden, Switzerland, Pakistan, UK	300	^1H, ^{13}C, ^{19}F	Dexamethasone 21-acetate, triamcinolone acetonide, unknown corticosteroid	[11]
	ACEi	China	500	^1H	1-[(2S,4R)-thio-2-methylpropionyl]-5-d-ethanedicarboxylicacid]-l-proline, 1-[(2S,4S)-thio-2-methylpropionyl-5-d-ethanedicarboxylicacid]-l-proline	[12]
	Sertraline HCl, Alprazolam, Vitamin D3, Enalapril Maleate	Brazil	500	^1H	Lower level of API	[13]
	Paracetamol	Brazil	500	^1H	Para-aminophenol	
	Omeprazole	Europe	600	^1H	Propanol, acetone, dichloromethane, dibutyl phthalate	[14]
	Cialis®	Germany, Syria, India, China		^1H-1H-2D-DOSY	No API, vardenafil, homosildenafil	[7]

(continued)

TABLE 7.1 (Continued)
Applications of qNMR spectroscopy for the analysis of counterfeit drugs

Technique/ Method	Drug	Location	NMR (MHz)	Nuclei/ Experiment	Adulterant	Ref
	mAbs	Switzerland	600	^1H	No API, hydroxyethyl starch, benzyl alcohol	[18]
	Methamphetamine	New Zealand	40	^1H, ^{31}P	Piperazionafil, Isopiperazionafil	[31]
	Grapefruit seed extract	Sweeden	300, 600	^1H, ^{13}C	Benzethonium	[30]
	Ibandronic acid, amantadine, ambroxol, Lercanidipine	Lab	400	PULCON, ^1H	No adulterants	[17]
Analytical	Miltefosine	Bangladesh	400	^1H, ^{31}P	No API	[32]
	Anabolic Steroids	Brazil	600	^1H	No API/ Less API/ peanut oil, benzyl alcohol	[10]
	Mix of herbs/sugars	South Europe	500	^1H	PP-THHS, THIO, SILD, THP, PHE, OST	[24]
	Anti-rheumatic	Vietnam	500	^1H, ^{13}C	Acetaminophen, sulfamethoxazole, indomethacin, trimethoprim	[23]
	DS	Germany	500	^1H, 2D-COSY	Piperazionafil, isopiperazionafil	[25]
	Weight-loss compounds	South Korea	600	^1H, ^{13}C	Levothyroxine, sennoside A & B, phenolphthaleine, unknown compounds	[26]
	Saposhnikoviae radix herbal supplements	Japan	600	^1H, ^{13}C	Xanthalin	[20]

	Sample	Country	Freq (MHz)	Analytes	Method	Ref
	Megaton	South Korea	600	Acetylvardenafil	¹H, ¹³C	[19]
	OTC herbal aphrodisiacs	US	500	thiohydroxyhomosildenafil	¹H, ¹³C	[21]
	Sexual enhancers	France	500	Sildenafil, tadalafil, vardenafil, flibanserin, testosterone, phentholamine, yohimbine, icariin	¹H	[27]
	Turmeric- containing DS	Italy	400	Syncumin	¹H	[28]
	DS	Germany	400	Sibutramine, mesterolone, oxymetholone, monacolin K, vincopocetine, evodiamine, caffeine, testosterone propionate, kavalactones, dehydro-epi-androsterone	¹H	[29]
Comparative	Anabolic Steroids	Brazil	600	No API/ Less API/ peanut oil, benzyl alcohol	¹H	[10]
Polarization Transfer	Ibuprofen, naproxen	Belgium, France, Portugal, UK, US	75, 100	Fingerprint of manufacture	¹³C, ²H, ¹H, INEPT	[15]
	Ibuprofen	US, UK, Belgium, France, Portugal	600	Fingerprint of manufacture	¹³C, DEPT, INEPT	[16]
TD	mAbs	Lab	20	None/less API	¹H	[8]

NMR: *nuclear magnetic resonance,* OTC: *over-the-counter,* ACEi: *angiotensin-converting-enzyme inhibitor,* API: *active pharmaceutical ingredient,* DOSY: *diffusion ordered spectroscopy,* mAbs-*monoclonal antibodies,* PULCON: *pulse length-based concentration,* PP-THHS: *propoxylphenyl-thiohydroxyhomosildenafil,* THIO: *thiosildenafil,* SILD: *sildenafil,* THP: *tetrahydropalmatine,* PHE: *phentolamine,* OST: *osthole,* INEPT: *intense nuclei enhanced polarization transfer,* DEPT: *distortionless enhancement by polarization transfer;* DS: *dietary supplements,* TD: *Time-domain*

Wilczyński *et al.,* utilized qualitative LF-¹H-NMR relaxometry to verify the authenticity of a 100mg sample of "Viagra" purchased on the black market in Poland, by comparing it to original Viagra. [35] The authors observed bi-exponential relaxation, hence two components were present in the counterfeit product, while the original Viagra had a single-exponential relaxation process. As a result, the authors concluded that LF-¹H-NMR relaxometry was useful to identify counterfeit drugs. [35]

Some other qualitative studies have utilized high-field (HF) ¹H-NMR to identify counterfeit drugs. Kesanakurti *et al.,* analyzed 24 samples of Sarsaparilla, a popular natural health product, purchased in Canada. [36] Included were one sample each of *Decalepis hamiltoni, Smilax officinalis,* and *Smilax oranata*; five of *Hemidesmus indicus*; seven each of *Smilax aris tolochiifolia* and *Pteridium aquilinum*; and two of *Sarsaparilla root.* The samples were analyzed in triplicate on a 600 MHz instrument. The authors were able to identify and group the samples into four taxonomically distant groups of Sarsaparilla. [36]

Rodomonte *et al.,* utilized HF-NMR on both ¹H and ¹³C nuclei on a 400 MHz instrument to analyze 16 in-lab synthesized analogs of sildenafil, thiosildenafil, and acetilsildenafil. [37] The synthesized samples included sildenafil, homosildenafil, hydroxysildenafil, dimethylsildenafil, piperidinosildenafil, pyrrolidinosildenafil, morpholinosildenafil, diethylaminosildenafil, thiosildenafil, thiohomosildenafil, thiohydroxyhomosildenafil, thiodimethylsildenafil, noracetilsildenafil, acetilsildenafil, hydroxyacetilsildenafil, dimethylacetilsildenafil, and morpholinoacetilsilddenafil. They aimed to create a reference for future analysis of counterfeit drugs. Using HF-NMR they found which signals of the samples were predictive of adulterants. Of the 16 synthetic samples, 14 were already reported in the literature. Pyrrolidinosildenafil was novel, while diethylaminosildenafil was previously cited, but never as an adulterant. [37]

Similarly, Hasegawa *et al.,* utilized ¹H- and ¹³C-NMR to identify the DS "khaki powder", marketed for tonic effect, purchased online. [38] They dissolved the sample in DMSO-d$_6$ and ran the NMR on an 800 MHz instrument. They were able to detect tadalafil in the sample which was further confirmed by HPLC and MS. [38]

Goda *et al.,* tested herbal supplements, by utilizing ¹H- and ¹³C-NMR on samples mixed dried plants purchased online in Japan. [39] They were able to detect the presence of a cannabimimetic phenyl-acetyl indole in the samples, which was confirmed utilizing MS. [39]

Additional experiments provided more clarity and identity of samples utilizing HF-2D-NMR. For example, Venâncio *et al.,* utilized 2D-DOSY-NMR to identify a sample of an anti-inflammatory drug sold illegally in Brazil. [40] Using a 400 MHz NMR, they found ranitidine and a mixture of orphenadrine citrate, piroxicam, and dexamethasone in the samples. Similarly, Fernández *et al.,* analyzed 16 tablets of artesunate, an antimalarial drug purchased in Cambodia utilizing 2D-DOSY-¹H-NMR on a 500 MHz instrument. [41] They determined that only six of the samples were genuine, while the remaining ten containing the wrong API, such as acetaminophen, and/or excipients were starch, lactose, sucrose, dextrin, or stearate. Additionally, Rebiere *et al.*, utilized 2D-NMR on a 500 MHz instrument to analyze sixteen suspicious samples of so-called Viagra obtained from India, Syria, and China. [42] They found adulterants such as, sildenafil citrate as the API as well as polyethylene glycol instead of lactose and triacetin as the excipients. [42] Likewise, Esteve-Turrillas *et al.*, utilized 2D-DOSY-NMR on a sample of ketamine purchased online. [43] They found the presence of 3',4'-methylenedioxy-2,2-dibromobutyrophenone, which they confirmed via IR and GC/MS. [43] Similarly, Göker *et al.*, utilized the same method to analyze two cans of 150 mL each of energy drinks purchased in Turkey. [44] They found the presence of propoxyphenyl which they verified using IR and MS. [44] Overall, the use of NMR to qualitatively identify counterfeits has been shown to be an effective, but costly, approach (see Table 7.2).

7.5 CONCLUSION

NMR spectroscopy is an effective approach for detecting and quantifying both the APIs and impurities in counterfeit drugs. Different methods, such as quantitative and qualitative NMR, have been utilized each with their own inherent advantages and disadvantages (see Table 7.3). Compared to other techniques, NMR is faster, nondestructive, and is able to identify and quantify APIs. Conversely, NMR is costly and often requires highly trained technicians to operate and interpret the data.

Of the quantitative NMR approaches, the direct determination method was effective as it did not require the need for reference standards. This is especially useful when these standards are not readily available, such as with new drugs. PULCON was particularly unique in not needing multiple calibrations of samples being analyzed, while TD-NMR did not need the sample to be dissolved in a deuterated solvent. Utilizing polarization transfer methods such as DEPT and INEPT, it was possible to determine a fingerprint for different manufacturers so that the origin of the drug could be identified. Despite these advantages, there are some setbacks to the quantitative analysis and NMR in general. NMR instruments are often big and expensive, as well as costly to maintain. Additionally, they can be complex to operate, and with HF-NMR require a cryogenic exchange.

Of the qualitative analysis, the benchtop and LF-NMR were effective for rapid screening to identify the components of counterfeit drugs. These studies utilize much simpler and less expensive NMR machines that require less space and complexity to operate. However, they are not as precise and may not be able to identify all compounds.

NMR is a powerful technology effective in identifying and quantifying potential counterfeit drugs, as well as creating a fingerprint to find their origin. Hence, it contributes to overcoming the illegal drug market and controlling the general public health threat.

TABLE 7.2
Applications of qualitative NMR spectroscopy for the analysis of counterfeit drugs

Technique/Method	Drug	Location	NMR (MHz)	Nuclei/Experiment	Adulterant	Ref
LF	Weight loss DS	Internet	60	^{1}H	Sibutramine, phenolphthalein, both	[33]
LF-relaxometry	Viagra	Poland	40	^{1}H	Sildenafil citrate	[35]
Benchtop LF	Sexual enhancement & weight-loss DS	Internet	60	^{1}H	Sildenafil, tadalafil	[34]
HF	Sarsaparilla	Canada	600	^{1}H	Distant taxonomical groups of Sarsaparilla	[36]
HF	Analogs of sildenafil, thiosildenafil, & acetilsildenafil	Lab	400	^{1}H, ^{13}C	Pyrrolidinosildenafil, diethylaminosildenafil	[37]
HF	"khaki powder" DS	Internet	800	^{1}H, ^{13}C	Tadalafil	[38]
HF	Herbal supplements	Japan	600, 150	^{1}H, ^{13}C	Phenyl-acetyl indole	[39]
HF	Anti-inflammatory	Brazil	400	^{1}H, 2D-DOSY	Ranitidine, orphenadrine citrate, piroxicam, dexamethasone	[40]
HF	Artesunate	Cambodia	500	^{1}H, 2D-DOSY	Acetaminophen, starch, lactose, sucrose, dextrin, stearate	[41]
HF	Viagra®	India, Syria, China	500	^{1}H, 2D-DOSY	Sildenafil citrate, polyethylene glycol	[42]
HF	Ketamine	Internet	300, 75	^{1}H, ^{13}C, DEPT, 2D-DOSY	3',4'-methylenedioxy-2,2-dibromobutyrophenone	[43]
HF	Energy drink	Turkey	400, 100	^{1}H, ^{13}C, 2D-DOSY	Propoxyphenyl	[44]

LF: *low-field*, HF: *high-field*, NMR: *nuclear magnetic resonance*, DS: *dietary supplements*, DEPT: *distortionless enhancement by polarization transfer*, DOSY: *diffusion ordered spectroscopy*, mAbs: *monoclonal antibodies*

TABLE 7.3
Advantages and disadvantages of qualitative and quantitative NMR

NMR Method	Advantages	Disadvantages
Quantitative	• Faster analysis compared to other techniques that require separation • Nondestructive • Signal intensity depends directly on the number of protons and their concentration • Identifies and quantifies API • Determination of concentration of components in mixtures • Direct determination NMR is effective when reference standards are not available • PULCON does not need multiple calibrations of samples • TD-NMR effective in more complex compounds and mixes of drugs in solid states; does not need a deuterated solvent • 2D gives more precise information about the components • DEPT and INEPT are useful in determining the fingerprint of manufacturers	• Expensive to maintain • Complex to operate • Requires cryogens • Larger in size; takes up space • Only applied to components with known chemical structure • Components need to be soluble in deuterated solvent • Often requires high resolution NMR machines
Qualitative	• Non-destructive • Often sufficient to determine if the API is present • Rapid screening, quick sample preparation, rapid spectral recording • HF-NMR provides a more precise structure of the molecules • Benchtop and LF-NMR are more simple and less expensive	• HF-NMR is more expensive and complex to operate • Benchtop and LF-NMR have less precision; may not be able to identify all compounds

HF: *high-field*, LF: *low-field*, DEPT: *distortionless enhancement by polarization transfer*, INEPT: *intense nuclei enhanced polarization transfer*, TD: *Time-domain*, PULCON: *pulse length-based concentration*, NMR: *nuclear magnetic resonance*

REFERENCES

1. Martino, R., Malet-Martino, M., Gilard, V., & Balayssac, S. (2010). Counterfeit drugs: analytical techniques for their identification. *Analytical and Bioanalytical Chemistry, 398*(1), 77–92. doi:10.1007/s00216-010-3748-y

2. Bolla, A. S., Patel, A. R., & Priefer, R. (2020). The silent development of counterfeit medications in developing countries – a systematic review of detection technologies. *International Journal of Pharmaceutics, 587*(119702), 119702. doi:10.1016/j.ijpharm.2020.119702

3. Holzgrabe, U., & Malet-Martino, M. (2011). Analytical challenges in drug counterfeiting and falsification – the NMR approach. *Journal of Pharmaceutical and Biomedical Analysis, 55*(4), 679–687. doi:10.1016/j.jpba.2010.12.017

4. Singh, S., Prasad, B., Savaliya, A., Shah, R., Gohil, V., & Kaur, A. (2009). Strategies for characterizing sildenafil, vardenafil, tadalafil and their analogues in herbal dietary supplements, and detecting counterfeit products containing these drugs. *Trends in Analytical Chemistry: TRAC, 28*(1), 13–28. doi:10.1016/j.trac.2008.09.004

5. Mackey, T. K., Cuomo, R., Guerra, C., & Liang, B. A. (2015). After counterfeit Avastin® – what have we learned and what can be done? *Nature Reviews. Clinical Oncology, 12*(5), 302–308. doi:10.1038/nrclinonc.2015.35

6. Assemat, G., Dubocq, F., Balayssac, S., Lamoureux, C., Malet-Martino, M., & Gilard, V. (2017). Screening of "spice" herbal mixtures: From high-field to low-field proton NMR. *Forensic Science International, 279*, 88–95. doi:10.1016/j.forsciint.2017.08.006

7. Vaysse, J., Gilard, V., Balayssac, S., Zedde, C., Martino, R., & Malet-Martino, M. (2012). Identification of a novel sildenafil analogue in an adulterated herbal supplement. *Journal of Pharmaceutical and Biomedical Analysis, 59*, 58–66. doi:10.1016/j.jpba.2011.10.001

8. Akhunzada, Z., Wu, Y., Haby, T., Jayawickrama, D., McGeorge, G., La Colla, M., … Abraham, A. (2021). Analysis of biopharmaceutical formulations by Time Domain Nuclear Magnetic Resonance (TD-NMR) spectroscopy: A potential method for detection of counterfeit biologic pharmaceuticals. *Journal of Pharmaceutical Sciences, 110*(7), 2765–2770. doi:10.1016/j.xphs.2021.03.011

9. Pindelska, E., Szeleszczuk, L., Pisklak, D. M., Mazurek, A., & Kolodziejski, W. (2015). Solid-state NMR as an effective method of polymorphic analysis: solid dosage forms of clopidogrel hydrogensulfate. *Journal of Pharmaceutical Sciences, 104*(1), 106–113. doi:10.1002/jps.24249

10. Ribeiro, M. V. de M., Boralle, N., Pezza, H. R., & Pezza, L. (2018). Authenticity assessment of anabolic androgenic steroids in counterfeit drugs by1H NMR. *Analytical Methods: Advancing Methods and Applications, 10*(10), 1140–1150. doi:10.1039/c8ay00158h

11. McEwen, I., Elmsjö, A., Lehnström, A., Hakkarainen, B., & Johansson, M. (2012). Screening of counterfeit corticosteroid in creams and ointments by NMR spectroscopy. *Journal of Pharmaceutical and Biomedical Analysis, 70*, 245–250. doi:10.1016/j.jpba.2012.07.005

12. Shen, S., Yang, X., & Shi, Y. (2015). Application of quantitative NMR for purity determination of standard ACE inhibitors. *Journal of Pharmaceutical and Biomedical Analysis, 114*, 190–199. doi:10.1016/j.jpba.2015.05.021

13. Dos Santos Ribeiro, H. S., Dagnino, D., & Schripsema, J. (2021). Rapid and accurate verification of drug identity, purity and quality by 1H-NMR using similarity calculations and differential NMR. *Journal of Pharmaceutical and Biomedical Analysis, 199*(114040), 114040. doi:10.1016/j.jpba.2021.114040

14. Rebiere, H., Grange, Y., Deconinck, E., Courselle, P., Acevska, J., Brezovska, K., … Bertrand, M. (2022). European fingerprint study on omeprazole drug substances using a multi analytical approach and chemometrics as a tool for the discrimination of manufacturing sources. *Journal of Pharmaceutical and Biomedical Analysis, 208*(114444), 114444. doi:10.1016/j.jpba.2021.114444

15. Remaud, G. S., Bussy, U., Lees, M., Thomas, F., Desmurs, J.-R., Jamin, E., … Akoka, S. (2013). NMR spectrometry isotopic fingerprinting: a tool for the manufacturer for tracking active pharmaceutical ingredients from starting materials to final medicines. *European Journal of Pharmaceutical Sciences: Official Journal of the European Federation for Pharmaceutical Sciences, 48*(3), 464–473. doi:10.1016/j.ejps.2012.12.009

16. Bussy, U., Thibaudeau, C., Thomas, F., Desmurs, J.-R., Jamin, E., Remaud, G. S., … Akoka, S. (2011). Isotopic finger-printing of active pharmaceutical ingredients by 13C NMR and polarization transfer techniques as a tool to fight against counterfeiting. *Talanta, 85*(4), 1909–1914. doi:10.1016/j.talanta.2011.07.022

17. Monakhova, Y. B., Kohl-Himmelseher, M., Kuballa, T., & Lachenmeier, D. W. (2014). Determination of the purity of pharmaceutical reference materials by 1H NMR using the standardless PULCON methodology. *Journal of Pharmaceutical and Biomedical Analysis, 100*, 381–386. doi:10.1016/j.jpba.2014.08.024

18. Casagrande, F., Dégardin, K., & Ross, A. (2020). Protein NMR of biologicals: analytical support for development and marketed products. *Journal of Biomolecular NMR, 74*(10–11), 657–671. doi:10.1007/s10858-020-00318-w

19. Lee, H.-M., Kim, C. S., Jang, Y. M., Kwon, S. W., & Lee, B.-J. (2011). Separation and structural elucidation of a novel analogue of vardenafil included as an adulterant in a dietary supplement by liquid chromatography-electrospray ionization mass spectrometry, infrared spectroscopy and nuclear magnetic resonance spectroscopy. *Journal of Pharmaceutical and Biomedical Analysis, 54*(3), 491–496. doi:10.1016/j.jpba.2010.09.022

20. Maruyama, T., Ezaki, M., Shiba, M., Yamaji, H., Yoshitomi, T., Kawano, N., … Kawahara, N. (2018). Botanical origin and chemical constituents of commercial Saposhnikoviae radix and its related crude drugs available in Shaanxi and the surrounding regions. *Journal of Natural Medicines, 72*(1), 267–273. doi:10.1007/s11418-017-1149-7

21. Reepmeyer, J. C., & D'Avignon, D. A. (2009). Use of a hydrolytic procedure and spectrometric methods in the structure elucidation of a thiocarbonyl analogue of sildenafil detected as an adulterant in an over-the-counter herbal aphrodisiac. *Journal of AOAC International, 92*(5), 1336–1342. doi:10.1093/jaoac/92.5.1336

22. Silvestre, V., Mboula, V. M., Jouitteau, C., Akoka, S., Robins, R. J., & Remaud, G. S. (2009). Isotopic 13C NMR spectrometry to assess counterfeiting of active pharmaceutical ingredients: site-specific 13C content of aspirin and paracetamol. *Journal of Pharmaceutical and Biomedical Analysis, 50*(3), 336–341. doi:10.1016/j.jpba.2009.04.030

23. Wiest, J., Schollmayer, C., Gresser, G., & Holzgrabe, U. (2014). Identification and quantitation of the ingredients in a counterfeit Vietnamese herbal medicine against

rheumatic diseases. *Journal of Pharmaceutical and Biomedical Analysis*, 97, 24–28. 10.1016/j.jpba.2014.04.013

24. Vaysse, J., Gilard, V., Balayssac, S., Zedde, C., Martino, R., & Malet-Martino, M. (2012). Identification of a novel sildenafil analogue in an adulterated herbal supplement. *Journal of Pharmaceutical and Biomedical Analysis*, 59, 58–66. doi:10.1016/j.jpba.2011.10.001

25. Wollein, U., Eisenreich, W., & Schramek, N. (2011). Identification of novel sildenafil-analogues in an adulterated herbal food supplement. *Journal of Pharmaceutical and Biomedical Analysis*, 56(4), 705–712. doi:10.1016/j.jpba.2011.07.012

26. Lee, J. H., Park, H. N., Kim, N. S., Park, S., Bogonda, G., Oh, K., & Kang, H. (2019). Application of screening methods for weight-loss compounds and identification of new impurities in counterfeit drugs. *Forensic Science International*, 303(109932), 109932. doi:10.1016/j.forsciint.2019.109932

27. Gilard, V., Balayssac, S., Tinaugus, A., Martins, N., Martino, R., & Malet-Martino, M. (2015). Detection, identification and quantification by 1H NMR of adulterants in 150 herbal dietary supplements marketed for improving sexual performance. *Journal of Pharmaceutical and Biomedical Analysis*, 102, 476–493. doi:10.1016/j.jpba.2014.10.011

28. Kim, S. B., Bisson, J., Friesen, J. B., Bucchini, L., Gafner, S., Lankin, D. C., … McAlpine, J. B. (2021). The untargeted capability of NMR helps recognizing nefarious adulteration in natural products. *Journal of Natural Products*, 84(3), 846–856. doi:10.1021/acs.jnatprod.0c01196

29. Monakhova, Y. B., Kuballa, T., Löbell-Behrends, S., Maixner, S., Kohl-Himmelseher, M., Ruge, W., & Lachenmeier, D. W. (2013). Standardless 1H NMR determination of pharmacologically active substances in dietary supplements and medicines that have been illegally traded over the internet: Standardless 1H NMR quantification of pharmacologically active substances. *Drug Testing and Analysis*, 5(6), 400–411. doi:10.1002/dta.1367

30. Bekiroglu, S., Myrberg, O., Ostman, K., Ek, M., Arvidsson, T., Rundlöf, T., & Hakkarainen, B. (2008). Validation of a quantitative NMR method for suspected counterfeit products exemplified on determination of benzethonium chloride in grapefruit seed extracts. *Journal of Pharmaceutical and Biomedical Analysis*, 47(4–5), 958–961. doi:10.1016/j.jpba.2008.03.021

31. Bogun, B., & Moore, S. (2017). 1 H and 31 P benchtop NMR of liquids and solids used in and/or produced during the manufacture of methamphetamine by the HI reduction of pseudoephedrine/ephedrine. *Forensic Science International*, 278, 68–77. doi:10.1016/j.forsciint.2017.06.026

32. Kaur, H., Seifert, K., Hawkes, G. E., Coumbarides, G. S., Alvar, J., & Croft, S. L. (2015). Chemical and bioassay techniques to authenticate quality of the anti-leishmanial drug miltefosine. *The American Journal of Tropical Medicine and Hygiene*, 92(6 Suppl), 31–38. doi:10.4269/ajtmh.14-0586

33. Wu, N., Balayssac, S., Danoun, S., Malet-Martino, M., & Gilard, V. (2020). Chemometric analysis of low-field 1H NMR spectra for unveiling adulteration of slimming dietary supplements by pharmaceutical compounds. *Molecules (Basel, Switzerland)*, 25(5), 1193. doi:10.3390/molecules25051193

34. Pagès, G., Gerdova, A., Williamson, D., Gilard, V., Martino, R., & Malet-Martino, M. (2014). Evaluation of a benchtop cryogen-free low-field ¹H NMR spectrometer for the analysis of sexual enhancement and weight loss dietary supplements

adulterated with pharmaceutical substances. *Analytical Chemistry*, *86*(23), 11897–11904. doi:10.1021/ac503699u

35. Wilczyńki, S., Petelenz, M., Florek-Wojciechowska, M., Kulesza, S., Brym, S., Błońska-Fajfrowska, B., & Kruk, D. (2017). Verification of the authenticity of drugs by means of NMR relaxometry – Viagra ® as an example. *Journal of Pharmaceutical and Biomedical Analysis*, *135*, 199–205. doi:10.1016/j.jpba.2016.12.018

36. Kesanakurti, P., Thirugnanasambandam, A., Ragupathy, S., & Newmaster, S. G. (2020). Genome skimming and NMR chemical fingerprinting provide quality assurance biotechnology to validate Sarsaparilla identity and purity. *Scientific Reports*, *10*(1), 19192. doi:10.1038/s41598-020-76073-7

37. Mustazza, C., Borioni, A., Rodomonte, A. L., Bartolomei, M., Antoniella, E., Di Martino, P., ... Gaudiano, M. C. (2014). Characterization of Sildenafil analogs by MS/MS and NMR: a guidance for detection and structure elucidation of phosphodiesterase-5 inhibitors. *Journal of Pharmaceutical and Biomedical Analysis*, *96*, 170–186. doi:10.1016/j.jpba.2014.03.038

38. Hasegawa, T., Saijo, M., Ishii, T., Nagata, T., Haishima, Y., Kawahara, N., & Goda, Y. (2008). Structural elucidation of a tadalafil analogue found in a dietary supplement. *Shokuhin Eiseigaku Zasshi. Journal of the Food Hygienic Society of Japan*, *49*(4), 311–315. doi:10.3358/shokueishi.49.311

39. Uchiyama, N., Kikura-Hanajiri, R., Kawahara, N., Haishima, Y., & Goda, Y. (2009). Identification of a cannabinoid analog as a new type of designer drug in a herbal product. *Chemical & Pharmaceutical Bulletin*, *57*(4), 439–441. doi:10.1248/cpb.57.439

40. Silva, L. M. A., Filho, E. G. A., Thomasi, S. S., Silva, B. F., Ferreira, A. G., & Venâncio, T. (2013). Use of diffusion-ordered NMR spectroscopy and HPLC-UV-SPE-NMR to identify undeclared synthetic drugs in medicines illegally sold as phytotherapies: DOSY experiment and HPLC-UV-SPE-NMR to identify the components and/or potential adulterations in a drug. *Magnetic Resonance in Chemistry: MRC*, *51*(9), 541–548. doi:10.1002/mrc.3984

41. Nyadong, L., Harris, G. A., Balayssac, S., Galhena, A. S., Malet-Martino, M., Martino, R., ... Gilard, V. (2009). Combining two-dimensional diffusion-ordered nuclear magnetic resonance spectroscopy, imaging desorption electrospray ionization mass spectrometry, and direct analysis in real-time mass spectrometry for the integral investigation of counterfeit pharmaceuticals. *Analytical Chemistry*, *81*(12). doi:10.1021/ac900384j

42. Rebiere, H., Guinot, P., Chauvey, D., & Brenier, C. (2017). Fighting falsified medicines: the analytical approach. *Journal of Pharmaceutical and Biomedical Analysis*, *142*, 286–306. doi:10.1016/j.jpba.2017.05.010

43. Armenta, S., Gil, C., Ventura, M., & Esteve-Turrillas, F. A. (2020). Unexpected identification and characterization of a cathinone precursor in the new psychoactive substance market: 3',4'-methylenedioxy-2,2-dibromobutyrophenone. *Forensic Science International*, *306*(110043), 110043. doi:10.1016/j.forsciint.2019.110043

44. Alp, M., Coşkun, M., & Göker, H. (2013). Isolation and identification of a new sildenafil analogue adulterated in energy drink: propoxyphenyl sildenafil. *Journal of Pharmaceutical and Biomedical Analysis*, *72*, 155–158. doi:10.1016/j.jpba.2012.09.017

Index

A

Abortion, illegal drugs 88
Acetilsildenafil 184, 186
Active pharmaceutical ingredient (API) 3, 7, 16,
    21, 41, 47, 49, 57, 79, 99, 100, 103, 121,
    124, 126, 133, 172
    analytical curve method 174
    fingerprinting 177
    legitimate pharmaceutical manufacturers 8
    monograph methods 18
    oral formulations 7
    poor quality/fake 8
    quantification of 48, 157
    rate of release 18
    substandard and falsified (SF) medical
        products 41, 42
    suspected product characterization 42–43
Alanoferm cream 175
Alprazolam 89, 120, 176
Alumina, adsorbent film 14
Aluminium crimping caps 160
Amantadine 178
Ambroxol 178
Amoxicillin 2, 20, 21, 25, 28, 79, 80, 112, 121,
    122, 147, 159
    authenticity of 121
    PAD reagent 20
    primary amine groups 21
    quantification of 159
Amphetamine 107, 109, 110, 180
Ampicillin 3, 79, 80, 112
    British Pharmacopeia monograph 3
    PAD reagent 20
Anabolic-androgenic steroids (AAS) 84, 107
    testosterone 85
    ultra-liquid chromatographytandem mass
        spectrometry (UHPLC-MS/MS) 85
Anabolic steroid standards 174
Analgesics 46
Androgen replacement therapy (ART) 84
Angiotensin-converting-enzyme (ACE)
    inhibitors 176
Anorexics 46, 48, 110
Antiaging peptides 49
Antibiotics 46, 79, 110
Antibody, immunoassays based on 18
Antidepressants 110, 146

Antidiabetic drug 46, 48, 89
    medicines 146
    multivariate PLS calibration models 158
Antiepileptic drugs 46
Antihistamines 146
Antihypertensive, multivariate PLS calibration
    models 158
Antimalarial drugs 79, 80, 110
Antimicrobial drugs, counterfeit/substandard 159
Anti-obesity drugs 48
Antiparasitics 79
Anxiolytics 110
Aphrodisiacs 46
Aristolochia fanghi 61
Artemether, antimalaria medicine 82
Artemether-lumefantrine (AL) formulations 82
Artesunate, antimalaria medicine 82
Arthritis 85, 123
Artificial neural networks (ANN) 142
Asian Development Bank 26
Aspirin 78, 171
ASSURED criteria 11
Asthma 85
Atenolol, NIR spectrum of 141
Attenuated total reflectance Fourier transform
    infrared (ATR-FTIR) spectroscopy 138,
        158, 159, 160
    chemometrics 140
    vs. near-infrared (NIR) spectroscopy
        139–140
    non-destructive technique 139
    powder-mixing efficiency 138
    principle of 138
    signal variations 138
Attenuated total reflectance (ATR) techniques
    138, 158
Authentication identification
    active pharmaceutical ingredients 100
    analysis of 99
    counterfeit drugs 100
    dispersive liquid-liquid microextraction
        (DLLME) 101–102
    dispersive solid-phase extraction (d-SPE) 103
    hollow fiber-based liquid-phase
        microextraction 102
    liquid phase extraction (LPE) 101
    microextraction techniques 101
    microwave assisted extraction (MAE) 102

pressurized liquid extraction (PLE) 101
quick, easy, cheap, effective, rugged, and safe
(QuEChERS) 103
sample preparation 100–101
solid phase extraction (SPE) 102–103
solid phase microextraction (SPME) 103
ultrasonic assisted extraction 102
Authorized sales outlets (AT) 79
Avastin 172, 173
Ayurvedic/herbal healthcare products
(AHP)
in India 86
LC-MS/time of flight (LC-MS/TOF) 86
Azithromycin 147
antimicrobial drug 155
corn starch 29
for COVID-19 28

B

Bai Fu Kang Unguent ointment, 175
Beer-Lambert's Law 121
Belgium Federal Agency for Medicinal and
Health Products 89
Bempong, D. K. 11
Benazepril hydrochloride 176
Benchtop cryogen-free LF-^{1}H-NMR 180
Benign prostatic hyperplasia (BPH) 123
Benzodiazepines 48
Benzylpenicillin 79
Benzyl-sibutramine 87
Biggs, K. B. 11
Big money-making 171
Black fever, in Bangladesh 82
Boldenone 85
Bradford 50
Brazil, dietary supplements 110
Brazilian National Health Surveillance Agency
174
British Pharmacopoeial standard 79

C

Cadramine-V cream 175
Caffeine 110
Capillary columns 107
Capillary electrophoresis (CE) 104
Captagon tablets 110
Captopril 176
Cardiovascular disease 133
Cardiovascular therapy medicines 146
CBEx Handheld Raman Spectrometer 28
Ceftriaxone 79, 80
Cefuroxime 79, 80

Centers for Disease Control and Prevention
(CDC) 77
Central nervous system 42
Cheaper methods 134
Chemical color tests 4, 13, 16
Chemical ionization (CI) 107, 108
Chemical shift signal 174
China, counterfeit medications 90
Chinese/Aryuvedic medicine 57
Chinese herbal medicines 89, 104
Chloramphenicol 112
Chloroquine 29
analysis of 28
corn starch 29
for COVID-19 28, 133
Chlorpheniramine maleate 89
Chromatography 14, 78
chromatographic fingerprints 58
chromatographic instrumentations 104
chromatographic separation techniques 50,
51, 104
polypeptide drugs quantification 55
Cialis® 146, 177
lifestyle medicines 146
tablets 84
^{13}C-INEPT-NMR, on naproxen samples 177
Ciprofloxacin 79, 80, 147, 155
counterfeits and imitations screening 147
PAD reagent 20, 28
PMS campaign 6
Classification and regression tree (CART) 145
Clavulanic acid 79, 80, 112, 147
Clinical forensic toxicology 78
Clopidogrel (Plavix®) 121
Cloxacillin 112
Coartem tablets, of starch 7, 8
Coca body butter 175
Collision induced dissociation (CID)
fragmentation 51
Color tests
pharmaceutical products 14
WHO monographs 13
Column chromatography 14
Compendial methods 3
Coomassie blue staining 50
Corn starch 29
Correlation in wavelength (CWS) 142
chemometric methods 155
PCA methods 155
Corticosteroids 85
Counterfeit drugs 100, 120, 172
detrimental effects 88–90
development of 119

for erectile dysfunction 82–85
global crisis 77
herbal products *see* Falsified herbal products
liquid chromatography-mass spectrometry
    (LC-MS) 78
for obesity 86–88
for pathogenic diseases 79–82
qNMR spectroscopy 181–183
for steroids 85–86
Counterfeit herbal drugs
chromatographic techniques 104
definition of 98
gas chromatography/mass spectrometry (GC/
    MS) 105–106, 109–110
GC detectors 106–107
HPLC *vs.* GC/MS 110
liquid chromatography/tandem mass
    spectrometry (LC/MS/MS) 109
liquid chromatography technique (LC) 105
spectrophotometric techniques 104
thin layer chromatography (TLC) 105
undeclared drug detection 103
Counterfeit medicines 132
anabolic steroids 110
for concerning drugs 90
herbal medicines 90, 110, 171, 172
IR spectroscopic *see* infrared (IR)
    spectroscopic analytical tools
method validation *see* Method validation
NMR spectroscopy *see* Nuclear magnetic
    resonance (NMR) spectroscopy
opioids 120
spectrometer instrumentation *see* Mass
    spectrometry (MS)
substandard *see* substandard medicines
ultraviolet spectroscopy *see also* Ultraviolet
    visible spectroscopy (UV-Vis)
Counterfeit opioids 120
COVID-19 pandemic 120
counterfeit medicines, trade of 132
test strips 18
vaccines 120
Crumbling/foul-smelling tablets 9
Cushing's syndrome 85
Cyanopropylphenyl dimethyl polysiloxane 107
Cytochrome c, via column chromatography 121

D

Data driven-soft independent modelling of class
    analogies (DD-SIMCA) 144
Decalepis hamiltoni 184
Deconinck, Eric 112
Democratic Republic of Congo (DRC) 82, 121

N-Desmethyl-sibutramine (DMS) 87
Dexamethasone 21 acetate 175
Diabetes mellitus 133
Diazepam 10, 89, 110
Didesmethyl-sibutramine (DDMS) 87
Dietary supplements 44, 48
    anabolic steroids 110
    herbal medications 172
Diethylaminosildenafil 184
Differential scanning calorimetry (DSC) 159
Diffusion-ordered spectroscopy (DOSY) 173
Diflorasone diacetate 175
Dihydroartemisinin, antimalaria medicine 82
Dihydrocodone 110
Dimethylacetilsildenafil 184
Dimethylsildenafil 184
Dimethyl sulfone ($DMSO_2$) 174
2,4-Dinitrofluorobenzene 175
Diode-array detector (DAD) 50
Diphenylhidramine 89
Dispersive liquid-liquid microextraction
    (DLLME) 101–102
Dispersive solid phase extraction (d-SPE)
    103
Disposable immunoassay 26
Distortionless enhancement by polarization
    transfer (DEPT) 177
Diuretics 110
DNA metabarcoding 57
Doping peptides 89
Doxycycline 112, 147
Drug counterfeiting 98
    dangers of 98
    detection of 109
Drug detection, undeclared 103
Drug packaging 99
Drug quality control 120
Drug regulators, approval of 23
Drug screening applications 19
Durateston analysis 159

E

Eczema 85
Egypt, counterfeit medications 90
Electromagnetic spectrum 135
Electron capture detector (ECD) 55, 106
Electronic reference to access *in vivo*
    concentrations (ERETIC) 178
Electron ionization (EI) 107, 108
Electron transfer dissociation (ETD) 55
Electrospray ionization (ESI) 55
Enalapril maleate 176
Epimedium spp. 59, 61

Epimedium spp. Leaves 59
  MS fingerprints of 60
Epitalon 89
Epitalon peptide 50
Erectile dysfunction (ED) 82
  counterfeit drugs 82–85
  drugs 112
  by using LC-MS 84
Erythromycin 79, 80
Erythropoietin (EPO) 49, 173
Euclidean distance 143, 144
European Directorate for Quality of Medicines
    (EDQM) 41
European Medicines Agency (EMA) 49
European Pharmacopoeia (Ph. Eur.) 42
Europe, drug regulatory agencies 3

F

Face, muscle spasms of 10
Fake Temgesic vials 106
Falsified herbal products
  adulteration *see* Herbal drugs, adulteration
  forensic toxicology and chemistry 98
  with GC/MS 98
  preparations 98
Famotidine, quantity of 123
Fenethylline 110
Fentanyl molecule 18
Field evaluation 28
Field screening test
  definition of 6
  field evaluation 28
  Lao-Oxford-Mahosot Hospital-Wellcome
    Trust Research Unit (LOMWRU)
    26–27
  performance evaluation 28
  for pharmaceuticals 25–26
  SFP, problem solving 6–7
  target application 28
  US pharmacopoeia technology review
    program 27–29
Flame ionization detector (FID) 106
Flame photometric detector 106
Flibanserin 179
Fluoxetine 110
  counterfeits/imitations screening 147
  hidden APIs 110
  quality of 124
Food and Drug Administration (FDA) 24
Food dyes, quantity of 125
Food quality testing 18
Forensic toxicology 98, 101
Fourier transform infrared (FTIR) 137

skilled personnel 138
spectroscopy 104, 157
Fused silica GC columns 105
Fuzzy C-mean 158

G

Gas chromatography (GC) 4, 14, 43, 105
  volatilization of 106
Gas chromatography-mass spectrometry
    (GC/MS) 98, 104, 105–106, 109–110
  GC detectors 106–107
Gastrointestinal symptoms 89
Gastrointestinal (GI) tract 18
Generalised European Official Medicines
    Networks (GEON) 42
  API fingerprint program 43
  European GEON network 45
  Illegal Medicines Working group 44
Gentamicin 79, 80
Georgia Tech 26
Ghana
  antibiotics, quality of 78, 79
  populations 79
Glibenclamide 89
Gliclazide 89
Glimepiride 89
Glipizide 89
Gliquidone 89
Global crisis 77
Global Pharma Health Fund (GPHF)-Minilab™
    28
Glucocorticoids 112
Glucose sensors 20
Glyburide, with sildenafil produces 122
Glycopeptides 52
Good manufacturing practices (GMP) 42
  API manufacturing sites 42
  pharmaceutical manufacturers 7
GPHF Minilabs™ 15, 16, 17, 27
Grapefruit seed extract (GSE) 180
Guidelines to Combat and Test Counterfeit
    Medicines 15

H

Halfan, in Southeast Asia 80
Headspace 100
Heart disease 86
Heat-sealed foil packaging 9
Hemidesmus indicus 184
Hemp body butter 175
Herbal drug
  adulteration *see* Herbal drugs, adulteration

medicines 99
preparations 98
traditional medicines 57
Herbal drugs, adulteration
methods of 99
preparations 99
reasons 99
reasons for 99
Herbal products *see* Falsified herbal products;
Herbal drugs, adulteration
Herbal supplements 171–172
Herbal weight loss drugs 110
Herschel, William 135
^{1}H-^{1}H-2D-DOSY-NMR spectroscopy 177
Hierarchical clustering analysis (HCA) 142, 176
Higher-energy C-trap dissociation (HCD) 45
High-income countries (HIC) 132
low- and middle- income countries 2
monograph methods 4
pharmaceutical supply chain integrity 132
High performance liquid chromatography 3, 4,
14, 50, 78, 104, 120, 133, 174
High performance thin layer chromatography
(HPTLC) 15, 105
High-pressure liquid chromatography (HPLC) 40
assay, quinine dihydrochloride 156
falsified medicines 41–42
vs. GC/MS 110
illegal preparations, containing
macromolecules 48–50
illegal preparations, containing small
molecules 41–42
overview of 40–41
suspected APIs, characterization of 42–43
suspected finished products, characterization
of 43–48
ultra-high pressure chromatography (UHPLC)
40
with UV/DAD 40
High resolution mass spectrometry (HRMS) 51
Hit quality index (HQI) 142
HIV/AIDS 133, 172
^{1}H-NMR spectroscopy 176
Hoang VD 123
Hollow fiber-based liquid-phase microextraction
102
Homosildenafil 184
Hormones/anabolics 46
HRAM fragmentation patterns 44
Human chorionic gonadotropin (hGC) 49
Human growth hormone (hGH) 49
Hydrophilic interaction liquid chromatography
(HILIC) 50

Hydrophobic compounds 78
Hydroxyacetilsildenafil 184
Hydroxychloroquine 29
corn starch 29
for COVID-19 28
Hydroxymethylfurfural 125
Hydroxysildenafil 184
Hypoglycemia 89, 122

I

Ibandronic acid 178
Ibrahim AM 123
Ibuprofen, quantity of 123
Ilex paraguariensis 61
Illegal drugs
chromatography, polypeptide drugs
quantification 55
LC-MS, for identification 50–55
packaging/visual inspection 49
peptide drugs, LC-MS for identification 49–52
polypeptide, by European regulatory 48
protein drugs, LC-MS for identification 49,
50, 52–55
Illicit street drugs (idPAD) 20, 24
Imidapril hydrochloride 176
Immunoaffinity purification 55
Immunoassay testing 18
Indian aphrodisiac ayurvedic/healthcare products
48
Inductivity coupled plasma (ICP) 107
Inert gasses 107
Inflammatory bowel disease 85
infrared (IR) spectroscopic analytical tools
attenuated total reflectance Fourier transform
infrared (ATR-FTIR) spectroscopy
138–139, 139–140
chemometric approaches 140–146
counterfeit/substandard medicines 148–154
COVID-19 pandemic 132
fundamentals of 135
highperformance liquid chromatography
(HPLC) 133
HIV/AIDS 133
mid-infrared (MIR) spectroscopy 134–138
near-infrared (NIR) spectroscopy 134–137,
139–140
patients safety 132
spectrophotometric techniques 104
spectroscopy (IR spectroscopy) 3, 4, 104, 134,
135
substandard medicines *see* Counterfeit
medicines; Substandard medicines

Instrument-free field screening technologies
  analytical metrics 11–12
  disintegration/dissolution testing 18
  glassware-based field color tests 13–14
  lateral flow immunoassay strips 18–19
  Merck-GPHF MinilabTM 15–18
  packaging inspection 12
  paper analytical device (PAD) 19–23
  paper microfluidics 19
  test development 20
  thin-layer chromatography 14–15
Insulins 49
Intellectual property 23
Intense nuclei enhanced polarization transfer
      (INEPT) 177
International Conference on Harmonization
      (ICH) 174
International nonproprietray names (INN)
      molecules 42
International Pharmacopeia (IP) 3
International units (IU) 3
Investigational Device Exemption 24
Ionization methods 107
Ions, separation of 108
Ion-trap 44
Isoniazid
  Minlab manual page 17
  PAD reagent 20
Isotab, heart patients 8

J

Jähnke, Richard 16
Japan, drug regulatory agencies 3

K

Kenya, paper analytical device 22
K fuzzy means 142
Kim, S. H. 48
k-Nearest neighbour (k-NN) 144, 158
Korean Ministry of Food and Drug Safety 88
Kovacs, S. 11

L

Lamotrigine, quality of 123–124
Lao-Oxford-Mahosot Hospital-Wellcome Trust
      Research Unit (LOMWRU) 26
  Android and iOS operating systems 25
  field screening test 26–27
  PAD 26
Leave one out cross validation (LOOCV) 143
Legitimate pharmaceutical manufacturers 8

Lercanidipine 178
Lieberman, Marya 16
Lifesaving medications 49
Life-threatening diseases 133
Limit of detection (LOD) 111
Limit of quantification (LOQ) 111
Linear discriminant analysis (LDA) 143
Lipitor, cholesterol-lowering medicine 156
Liquid chromatography (LC) 40, 41, 43, 44, 50,
      105, 109
  LC-UV chromatography 61
  solvent, polarity index 78
Liquid chromatography-diode array detector
      (LC-DAD) methods 110
Liquid chromatography-mass spectrometry
      (LC-MS) 4, 43, 78
  counterfeit drugs, global crisis 78
  counterfeit medications, for pathogenic
      diseases 81
  counterfeit medications, for steroids 86
  coupled with mass spectrometry 78–79
  ESI-MS, corticosteroids 85
  screening methods 46
  tandem mass spectrometry (LC/MS/MS) 104,
      109
Liquid-liquid extraction (LLE) 101
Liquid phase extraction (LPE) 101
  disadvantages of 101
  organic solvents 101
Lisinopril 176
Low-and middle-income countries (LMICs) 2
  regulatory activities 4–6
  SFPs, prevalence of 2
Low-field ^{1}H-NMR (LF ^{1}H-NMR) 173
Low-income countries (LIC) 132
Low to middle-income countries (LMIC) 132
  enforcement activities 6
  field screening technologies, for
      pharmaceuticals 11
  screening methods 6

M

Magnesium sulfate (MgSO$_4$) 103
Mahalanobis distances 144
Malaria
  death rates 82
  RDTs 18
  in Southeast Asia 80
Martins AR 125
Mass spectrometry (MS) 40, 49, 50, 79, 106,
      120, 172
  chemical ionization (CI) 108

computer 109
detectors 107, 109
electron ionization (EI) 108
fingerprints, of *Epimedium spp. Leaves* 60
ion source 107–108
mass analyzer 108–109
mass-to-charge ratio (m/z) 79
Medicines detection 46
Melanotan II 89
Meningitis 11
Meridia 87
Metformin 89
  hypoglycemic drugs 89
  PAD reagent 20
Methadone 110
Methamphetamine 110
Method validation 110
  accuracy 111
  extraction efficiency, recovery of 111–112
  limit of detection (LOD) 111
  limit of quantification (LOQ) 111
  linearity 111
  precision 111
  selectivity 111
3',4'-Methylenedioxy-2,2-
  dibromobutyrophenone 185
Methylprednisolone 175
Metronidazole 79, 80, 122
Metronidazole tablet, classification of 155
Miafrid pure night cream 175
Microcrystalline cellulose, adsorbent film 14
Microextraction techniques 101
MicroPHAZIR RX 27
Micro total analytical systems (µTAS) 19
Microwave assisted extraction (MAE) 101, 102
Mid-infrared (MIR) spectroscopy 134
  molecular vibration, electromagnetic energy
    of 137
  vs. NIR 140
Mikami, E. 48
Miltefos 82
Minilab™
  chromatographic technique 14
  color tests 13
  limitation of 18
  semi-quantitative analysis 17
Minlab manual page 17
Mometasone furoate 175
Monoclonal antibodies (mAbs) 18, 49, 173
Monograph assays 3, 4
  preparation of 4
  quantitative monograph methods 4
Morpholinoacetilsilddenafil 184

Morpholinosildenafil 184
Multi-linear regression (MLR) analysis 145

N

Nandrolone decanoate (NDE) 173
Nano liquid chromatography (nanoLC) 51
National drug regulatory agencies 2
National medicines regulatory authority
  (NMRA) 41
Natural drugs 99
Near-infrared (NIR) spectroscopy 104, 134,
  155
  vs. attenuated total reflectance Fourier
    transform infrared (ATR-FTIR)
    spectroscopy 139–140
  chemical imaging (NIR-CI) 157
  disadvantage of 137
  multivariate classification models 157
  PLS-R calibration model 157
  qualitative 186
  radiation 136
  schematic of 136
Nitrogen phosphorous detector (NPD) 106
Non-sterile environments 119
Noracetilsildenafil 184
Nuclear magnetic resonance (NMR)
  spectroscopy 44, 104, 120, 172
  active pharmaceutical ingredient (API) 172
  "big money-making" industry 171
  counterfeit drugs, analysis of 181–183, 186
  counterfeit medication *see* Counterfeit
    medicines
  diffusion-ordered spectroscopy (DOSY) 173
  electromagnetic radiation 172
  high-field ¹H-NMR 173
  qualitative 180–185, 187
  quantitative 173–180, 187
  solid-state NMR (ssNMR) 173
  time domain NMR (TD-NMR) 173

O

Obesity
  counterfeit drugs 86–88
  global chronic disease 86–88
Oflaxacin 147
OMCL network 50
Open Science movement 25
Opioids
  abortion-inducing drugs 88
  substitution therapy 110
Optimal drug extraction 100
Osthole (OST) 179

Over-the-counter (OTC) 175
Oxandrolone (OXA) 173
Oxycodone 110

P

Packaging
  authentic products 10
  fake 12
  field screening 12
  illegal drugs 49
  inks/glues/heat-sealable foil 9
  inspection 12
  pharmaceutical 9
  physical packaging inspection 9
  security features 12
Pakistan, pharmaceutical manufacturer 8
Pandey A 122
Paper analytical device (PAD) 19–23
  active pharmaceutical ingredients 21
  Android and iOS operating systems 25
  antimalarial drugs 20
  automation of 25
  case studies 23–25
  chemotherapy drugs (chemoPAD) 20
  chromatography paper 19
  corn starch 29
  data analysis approach 23
  data analytics 21–23
  fringe diagram 21
  glassware, translation of 20
  hand stamping tool 22
  hydrophilic paper substrates 19
  illicit street drugs (idPAD) 20
  intellectual property 23
  in Kenya 22
  LOMWRU team 26
  mid-infrared spectrometer 27
  paper analytical devices making 20–21
  paper microfluidics 19, 20
  performance evaluation 28
  pharmaceuctical dosage forms 19
  pharmaceutical FDFs 28
  pharmaceutical screening 19
  pilot-scale manufacture 24
  QA/QC procedures 24
  regulatory approval 23–25
  single-use 27
  test development 20
  working process 19
Paper microfluidic devices, laboratory analysis
    methods 20
Paracetamol 89, 146

chronic pain 122–123
multivariate PLS calibration models 158
Parikh P 122
Partial least squares-discriminant analysis
    (PLS-DA) 144
Partial least squares regression (PLS-R) 145
Pausinystalia yohimbe bark 59, 61
Pegasys syringes 171
Penicillin G 80
Peptide drugs 49
  illegal preparations, containing
    macromolecules 50
  through illicit internet pharmacies 53–54
Pharmaceutical quality
  definition of 3
  field screening technologies 7
Pharmaceuticals
  badly packaged/transported/stored 9–10
  dispensing 10–11
  lethal contempt 7–9
  packaging 9
  quality control purposes 11
PharmaChk device 18, 23
Pharmacopeial organizations 3
Phenformin 89
Phenolphthalein 180
Phenoxymethylpenicillin 79
Phentolamine (PHE) 179
Phone app/PADreader 28
Phosphodiesterase type-5 inhibitors (PDE-5Is)
    42, 46, 83
  ayurvedic/healthcare products 48
  counterfeit/illicit 83
  dietary supplements, by using LC-MS/MS 83
  dietary supplements/herbal products/sexual
    enhancers 83
  LC-MS screening methods 46
  therapeutic categories 46
Piperidinosildenafil 184
Plant food supplements 57
  herbal adulteration of 59
  "return to nature" 55
  traditional herbal medicines 55–61
Plasmodium falciparum malaria 82
Pneumonia 172
Polyacrylamide gel electrophoresis 50
Polyclonal antibodies 18
Polypeptide drugs 41
  by European regulatory 48
  quantification 55
Post-market surveillance (PMS) 2, 5
  enforcement activities 6
  laboratory assay 6

prioritize regulatory response 5
registration and import testing 5
type of 5
Potassium bromide (KBr) matrix 138
Pressurized liquid extraction (PLE) 101
Principal component analysis (PCA) 142, 176
    chemometric method 147
    clustering pattern 142
    discriminating tendencies 143
    use of 43
Principal component regression (PCR) 145
Probabilistic neural networks (PNN) 142
Propoxylphenylthiohydroxyhomosildenafil
    (PP-THHS) 179
Protected denomination of origin (PDO) 124
Protein drugs 49
    illegal preparations, containing
        macromolecules 50
    through illicit internet pharmacies 56
Psoriasis 85
Psychotropics 47
Pulse length-based concentration determination
    (PULCON) 178
Purge 100
Pyrazinamide 17
Pyrimethamine, heart patients 8
Pyrrolidinosildenafil 184

Q

QR codes 21
Quadratic discriminant analysis (QDA) 143
Quadrupole 44
Quadrupole mass analyzers 109
Quadrupole time of flight (Q-TOF) mass
    spectrometry
    electroscopy ionization 109
    ultra- high performance liquid
        chromatography 103
Quality assurance (QA) 18
Quality control (QC) 18
Quick, easy, cheap, effective, rugged, and safe
    (QuEChERS) 103
Quinapril hydrochloride 176
Quinine 122
Quinine dihydrochloride, HPLC assay 156

R

Rabies vaccines 173
Raman chemical imaging 156
Raman spectrometers 26
Raman spectroscopy 104, 156
Reductil 87

Relative standard deviation (RSD) 175
Repackaging, of pharmaceuticals 10
Repaglinide 89
Retention time (RT) 78
Reversed phase liquid chromatography 52
Rhinitis 85
Rifampicin 17
Ríos-Reina R 124
Rizatriptan 110
Rodionova, Oxana 157
Rosiglitazone 89
Roth, L. 7, 11

S

Sandwich immunoassay, uses 18
Sarsaparilla root 184
Scanning electron microscopy (SEM) 138
Selective androgen receptor modulators
    (SARMs) 42
Shenzhen 999 cream 175
Sibutramine
    anti-obesity drug 87
    mesterolone 180
Sildenafil (SILD) 110, 122, 179, 180, 184
Silhouette index 142
Silica, adsorbent film 14
Skin allergies 89
Skin tanning peptide 89
Smartphone applications 132
Smilax officinalis 184
Smilax oranata 184
Social media platforms 132
Sodium dodecyl sulphate 50
Soft independent modeling of class analogies
    (SIMCA) 61, 143, 144
    Data driven-soft independent modelling of
        class analogies (DD-SIMCA) 144
    PCA/SIMCA multivariate modelling 147
    supervised SIMCA chemometric technique 147
Software, for automated analysis 19
Solid phase extraction (SPE) 101, 102–103
Solid phase microextraction (SPME) 103
Solid-state NMR (ssNMR) 173
Sorbent, low consumption of 103
South Korea, counterfeit medications 90
Spectrophotometric techniques 104
Standard error of calibration (SEC) 145
Standard normal variate (SNV) 142
Stanozolol 173, 174
Steroids
    counterfeit drugs 85–86
    counterfeit medications 86

Stroke 86
Substandard and falsified (SF) medical products 40, 41
Substandard and falsified pharmaceuticals (SFPs) 2, 40
    chemical/spectroscopic screening technologies 7
    cost-effectiveness 27
    field screening, problem solving 6–7
    low-and middle-income countries (LMICs) 2
    problem of 4–6
    regulatory agency 7
    WHO survey 2
Substandard medicines
    infrared spectroscopy methods 148–154
    mid-infrared spectroscopy applications 157–160
    near-infrared (NIR) 146
    near-infrared spectroscopy applications 147–157
Sub-therapeutic antiinfective drugs 79
Sulphamethoxazole/trimethoprim 79, 80
Supervised chemometric methods 143
Supper tramadol-X 89
Support vector machines (SVM) 142
Sympathomimetics 110
Synthetic hypoglycemic drugs 89

T

Tadalafil 180
trans-Tadalafil 84
Tamol-X 89
Tanzanian Food and Drug Authority (TFDA) 25
Target application 28
Technology Review Program (TRP) 28
    target application 28
    USP Program 26–29
Tee-doll 89
Terazosin, benign prostatic hyperplasia 123
Testosterone 85
Testosterone propionate (TPR) 173
    Android and iOS operating systems 25
    four injectables 173
Tetracycline 79, 80
Tetrahydropalmatine (THP) 179
Therapeutic drug monitoring 11, 78
Thermo-desorption 100
Thin fused silica capillaries 107
Thin layer chromatography (TLC) 4, 14, 15, 40, 104, 105, 120
    advantages of 40
    liquid chromatography 40
Thiodimethylsildenafil 184

Thiohomosildenafil 184
Thiohydroxyhomosildenafil 184
Thiosildenafil (THIO) 179, 184
Time domain NMR (TD-NMR) 173
Time-of-Flight (ToF) 44
    MS technologies 44
    quadrupole (Q-ToF) 44
Titration-based assay method 3
Tolbutamide 89
p-Toluene sulfonic acid 20
Tonic effect 184
Toxic plants, in traditional medicinal products 55–61
Tramadol 110
    abortion-inducing drugs 88
    for anxiety 88
    for depression 88
    for premature ejaculation 88
Tramadol-X 89
Tramajack 89
Triamcinolone acetonide 175
Tribulus terrestris 59, 61
Trimethoprim 80
Triturations 58
Tuberculosis 133
Type 2 diabetes 86

U

Ugandan national drug authority's lab 28
Ultra-high performance liquid chromatography systems (UHPLC) 50
Ultra-high pressure chromatography (UHPLC) 40
Ultra-liquid chromatographytandem mass spectrometry (UHPLC-MS/MS) 85
Ultra-performance liquid chromatography (UPLC) 109
Ultra-performance liquid chromatography-mass spectrometry (UPLC-MS) 158
Ultrasonic assisted extraction 102
Ultraviolet (UV)
    detection 40, 133
    for Epimedium spp. Leaves 59
    fingerprints 59, 61
    lamp 16
    non-UV absorbent molecule 47
    visible (UV-Vis) see Ultraviolet visible spectroscopy (UV-Vis)
Ultraviolet visible spectroscopy (UV-Vis) 4, 104, 120–121
    anti-platelet medication 121
    counterfeit opioids 120
    qualitative analysis 121, 124–125

quantitative analysis 121–124
regulatory agencies 120
safety regulations 119
spectrophotometers 14
wavelength 121
Unauthorized sales outlets (UAT) 79
Unsupervised chemometric methods 142
US drug regulatory agencies 3
US FDA Forensic Chemistry Center 83
US Food and Drug Administration (FDA) 120
US Pharmacopeial Convention (USP) 3, 42
PQM+ program 24
USP Technology Review Program 26

V

Vaccines, COVID-19 pandemic 120
Vardenafil 179
Venlafaxine 110
Viagra 84, 146, 184, 185
by HPLC- DAD- MS/ MS 85
IR devices 146
phosphodiesterase inhibitors 105
PLS analysis 156
Vickers, S. 11

Visceral leishmaniasis 82
vitamin D3 176
Volumetric glassware 14

W

Weight reducing drugs 46
Whisky brands 125
Wi-Fi connection 25
World Health Organization (WHO) 41, 98, 132, 171
meta-analysis 2
monographs, on chemical color tests 13
WHO Essential Medicines List 7
WHO Medical Product Alerts 17

X

Xerox ColorQube wax printer 21
XM™-FOCUS cream 175
X-ray powder diffractometry (XRPD) 176

Z

Zaman group 18
Z™-skin repair cream 175

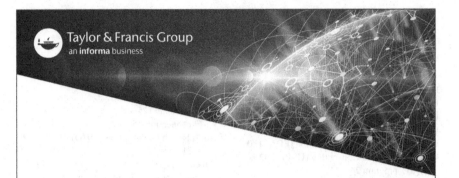

Printed in the United States
by Baker & Taylor Publisher Services